Carlson

JOURNEY THROUGH
SPACE

JOURNEY THROUGH

SPACE

TIM FURNISS

GALLERY BOOKS
An Imprint of W. H. Smith Publishers Inc.
112 Madison Avenue
New York City 10016

First published in the United States in 1991 by Gallery Books,
an imprint of W.H. Smith Publishers, Inc.,
112 Madison Avenue, New York, New York 10016
By arrangement with Reed International Books
Michelin House, 81 Fulham Road, London SW3 6RB

ISBN 0-8317-5270-X

Printed in Italy

Gallery Books are available for bulk purchase for sales promotions
and premium use. For details write or telephone the Manager of
Special Sales, W.H. Smith Publishers, Inc., 112 Madison Avenue,
New York, New York 10016. (212) 532-6600

CONTENTS

THE UNIVERSE

About 5,000 million people live on the planet we call Earth. This planet is one of nine that travel around (orbit) a star. This star is called the Sun. In addition to the planets, there are also moons, asteroids, comets, and other material, all of which are orbiting the Sun. This is the Solar System.

The Sun is a very small and rather insignificant star. It is one of over 100,000 million stars that swirl around in a whirlpool-shaped system known as a galaxy. Our Galaxy is called the Milky Way and is just one of many thousands of galaxies in the Universe.

Distances in Space

The scale of the Universe is astounding – it contains everything that exists. One way to understand distance within the Universe is to relate it to how long light takes to travel between points within it. Light travels at 186,282 mi/sec. This is called the speed of light. Light takes 1.26 seconds to travel between the Moon (our nearest neighbor in space) and Earth. This light actually comes from the Sun and is reflected off the surface of the Moon.

It takes 8 minutes 17 seconds for light to travel from the Sun to the Earth, so if the Sun were to go out, we would not notice for over eight minutes. The furthest known planet in our Solar System is called Pluto. The very faint light from the Sun that is reflected from Pluto takes 5 hours 20 minutes to travel to the Earth. Light from Proxima Centauri, the nearest star to the Sun, takes 4.22 years to reach Earth.

Our Solar System is situated on the edge of the Milky Way Galaxy. Light from the center of the Milky Way takes 27,700 years to reach us. Light from the other side of the Milky Way takes 62,700 years to reach the Earth. The nearest galaxy to the Milky Way is a jumble of stars called the Large Magellanic Cloud, and light from this takes 174,000 years to reach the Earth. The nearest galaxy that resembles the Milky Way is the Andromeda galaxy. Light from this takes 2,200,000 years to reach us. Light from the furthest reaches of the Universe that can be observed takes 14,000,000,000 years to reach the Earth.

Another way of looking at these huge distances is that although it takes just 1.26 seconds for light to travel between the Moon and the Earth, it took men three days to get there.

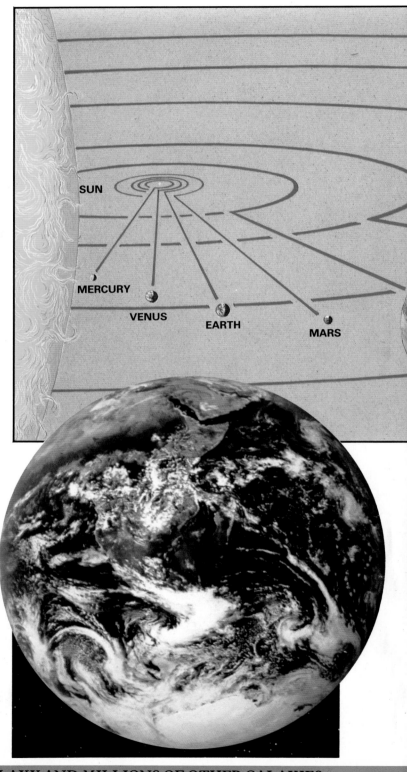

SUN
MERCURY
VENUS
EARTH
MARS

THE UNIVERSE IS EVERYTHING IN OUR GALAXY AND MILLIONS OF OTHER GALAXIES

PLUTO

NEPTUNE

URANUS

SATURN

JUPITER

Above: The Solar System consists of nine known planets that circle a small star called the Sun.

Left: The Earth is the third planet from the Sun, which is just one of billions of stars in the Milky Way. There are millions of other galaxies in the Universe.

Right: The Andromeda galaxy - the most distant object visible to the naked eye - is 2.2 million light years away.

1 LIGHT YEAR = 5,878,497,700,000 MI, THE DISTANCE LIGHT TRAVELS IN 1 YEAR

THE MILKY WAY

On a clear night, once your eyes have accustomed themselves to the darkness, you can see thousands of stars. Some bright, some quite faint. In fact, about 5,000 stars can be seen with the naked eye. The brightest star in the sky is called Sirius; it is 8.65 light years away. We can also see an area in the Orion constellation where stars are still being formed. It contains a massive red star, Betelgeuse, which is 310 light years away but is still the tenth brightest in the sky.

After a few minutes staring into the blackness sprinkled with these speckles of light, you will gradually notice a trail of what looks like faint cloud crossing directly overhead. With a pair of binoculars, it will become clear that this apparent cloud is a mass of thousands of stars. You are looking at the Milky Way. Our Galaxy is called a spiral galaxy because it is

Above: Our Milky Way Galaxy. Most of the galaxies visible in space are spirals like this.

Right: The Horsehead Nebula in the constellation of Orion, which can just be seen with the naked eye. This interstellar gas cloud is a birthplace of new stars in the Milky Way.

THE MILKY WAY GALAXY IS 100,000 LIGHT YEARS ACROSS, AND 20,000 LIGHT YEARS THICK

OTHER GALAXIES

There could be 100,000 million galaxies beyond our own. The closest galaxy that is spiral in shape is the Andromeda galaxy, 2.2 million light years away. Some spiral galaxies have a bar of stars across the center; these are called barred spirals. Elliptical galaxies are shaped like round balls or ovals. Astronomers believe that these consist of very old stars. Some of the biggest galaxies are elliptical ones; they can contain over a million million stars. Irregular galaxies have no shape at all. These are usually smaller than spiral galaxies and contain both young and old stars. They include the Large and Small Magellanic Clouds, which are visible in the southern hemisphere. They are the nearest galaxies to our own Galaxy.

like a spinning firework. From the side, the Milky Way looks relatively thin, rather like a disk. This disk is about 100,000 light years across, but just 20,000 light years thick. The Milky Way contains over 100,000 million stars. Situated toward the edge of the Galaxy, about 30,000 light years from the center is a small star – our Sun. The Sun lies in one of the many spiral "arms" of the Milky Way. Radio telescope and optical telescope observations have helped astronomers to reproduce what star groups they think are in the "arms" close to the Sun. The arm next closest to the center of the Milky Way contains the stars that form the constellations Perseus, Cassiopeia and Gemini which can be seen in the northern hemisphere. This is called the Perseus Arm. Then comes the Orion Arm, in which the Sun is situated, together with stars which make up much of the constellation of Orion and Cygnus. The outer arm contains stars which make up the most prominent southern hemisphere constellations, Sagittarius and Crux, the Southern Cross. It is called the Sagittarius Arm.

Three Regions of the Milky Way

The Milky Way consists of three regions: the "halo" outer region which surrounds the Galaxy, the "nucleus", a bulging central mass, and the spiral "disk".

The halo contains the oldest stars in the Galaxy which were formed within a vast cloud of gas over 10,000 million years ago. As the stars formed and the cloud began to rotate into a spiral galaxy, most collapsed toward the central nucleus but many of the older stars remained in a halo. Most are found in globular clusters.

The nucleus is a mass of stars. This bulging core measures about 20,000 by 10,000 light years. It consists of over 5 million formed stars and stars which are forming in the gas.

The disk region contains the youngest stars and clouds of hydrogen in which stars have yet to be formed. These regions have temperatures of over 18,000 °F and are called nebulas.

The Milky Way is slowly rotating, with the outer parts moving faster than the inner regions, so the Sun, on an outer arm of the spiral, moves at about 170 mi/sec. It takes about 250 million years for the Sun to circle the Galaxy, a period known as a cosmic year.

IT CONTAINS OVER 100,000 MILLION STARS • SUN IS 30,000 LIGHT YEARS FROM CENTER

BIG BANG

BIG BANG

GALAXIES FORM AND MOVE APART

UNIVERSE EXPANDS

IN BIG BANG THEORY, THE UNIVERSE WAS CREATED ABOUT 18,000 MILLION YEARS AGO

Left: Astronomers think that about 18,000 million years ago, a single, hot, dense mass of incredible energy suddenly exploded, beginning the expansion of the Universe.

Right: This is an American satellite, called Cobe, which was launched in 1990 to detect the radiation that was first emitted by the Big Bang.

How was the Universe created? This is one of the biggest puzzles to those who do not accept that God made it! In 1929, the astronomer Edwin Hubble made a discovery which revolutionized our understanding of the Universe, and led to modern theories of how it was born. By measuring the movement of distant galaxies, he found that the Universe appears to be expanding like a balloon being blown up. The galaxies are moving at incredible speeds, outward, away from an apparent center. For this reason, many astronomers or cosmologists believe that the Universe was formed by an immense explosion. This was an explosion not just of matter but also of time and space; the ultimate creation from nothing. It is called the Big Bang.

By measuring the rate at which the Universe is expanding, astronomers estimate that Big Bang happened about 18,000 million years ago. A single point of matter, space, and time in a hot, dense mass of incredible energy and heat suddenly exploded. Matter, time and space began to expand.

Matter, which was originally at a temperature of millions and millions of degrees, started to cool. Millions of years later, this cooling allowed atomic nuclei of helium gas to form, starting the process that would lead to the formation of clouds, in which stars could form.

Will the Universe Expand for Ever?

Some cosmologists who believe in the Steady State theory say that there has always been an expanding Universe, and that it will continue for ever as matter is continually being created. Others say that when the Big Bang finally ends and the Universe stops expanding, it will collapse back to another Big Bang, so the Universe is continually expanding and contracting. However, there is no evidence that the expansion of the Universe is now slowing down. According to another theory, the mechanism which created our Universe may well produce other universes, so that infinite space may be filled with several universe "islands".

ACCORDING TO STEADY STATE THEORY, THERE HAS ALWAYS BEEN A UNIVERSE

EARLY ASTRONOMY

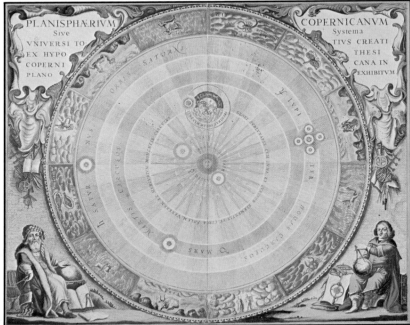

Left: The Polish astronomer Nicolaus Copernicus astounded the world when he realized that the Earth was not at the center of the Universe, but was an ordinary planet revolving around the Sun. His famous theory published in 1543 was called *The Revolutions of the Celestial Spheres*.

When people first started studying the sky, they believed that the Earth was at the center of the Universe, and flat, and that it did not move, but everything moved over it.

Early Chinese astronomers, about 3000 BC, realized that after 365 days the stars appeared to be in the same place as they were at day one. They called this period a "year", and even developed a calendar. But this scientific realization did not stop them from thinking that a solar eclipse – the Moon moving in front of the Sun – was a dragon eating the Sun!

The Egyptians made further astronomical discoveries. They even recognized that there was a Pole Star to which the Earth's axis points.

Rise of Modern Astrology

The early Greek astronomers charted the stars in detail. In the fourth century BC, the Greek philosopher Aristotle realized that the Earth was not flat, but round. He said that the stars and planets were stuck to the inside of a great rotating sphere with the Earth fixed at the center. The Greek astronomer Ptolemy, in the second century AD, wrote the *Almagest* which put the Earth at the center of the Universe, around which the other planets revolved.

ARISTOTLE (384–322BC) • PTOLEMY (100–175) • COPERNICUS (1473–1543) • GALILEO (1564–1642)

Left: The Italian astronomer Galileo Galilei was probably the first person to observe the night sky with a telescope, in 1609. He saw the phases of Venus, mountains and craters on the Moon, the moons of Jupiter, and realized that the Milky Way was made up of thousands of stars. His observations proved that the theories of Copernicus were correct.

It was not until the sixteenth century that the Polish astronomer Copernicus put forward the revolutionary idea that the Sun was at the center of the Universe and that the planets, including Earth, moved in circles around it. This theory caused consternation among many people at the time, including the Church authorities.

The German astronomer and mathematician Kepler studied the movement of planets around the Sun, and he realized that the planets move in ellipses (a shape like a squashed circle).

At the same time as Kepler was studying the motion of planets, the telescope was just coming into use. In 1609 the Italian scientist Galileo made a telescope and looked through it at the Moon and at the planet Jupiter. He saw craters on the Moon and moons orbiting Jupiter. The telescope revolutionized astronomy and led to many important discoveries; astronomers could now make accurate and detailed observations of the Universe. It was not until 1957, when the space age began, that a new method of discovery was possible.

COPERNICUS' THEORY PUBLISHED 1543 • GALILEO'S TELESCOPE INVENTED 1609

TELESCOPES

Telescopes are essential to astronomy. The Italian scientist Galileo was the first known person to use one to study the sky, in 1609. This was a refracting type, which used lenses to magnify an image. In 1668, Sir Isaac Newton built the first reflecting telescope, which used curved mirrors instead of lenses. In the late eighteenth century, the English observer William Herschel began building giant reflecting telescopes with which he made many discoveries, including the planet Uranus in 1781.

Today, observers have a vast array of instruments available to them. These include huge telescopes. The world's largest reflector is sited on Mount Semirodniki in the Caucasus Mountains, USSR. Its mirror is 20 ft in diameter, and is so sensitive it can detect the light from a candle 15,000 mi away. In addition to the visible light – that is what we can see with our human eyes – astronomers can also observe the Universe in other areas of the electromagnetic spectrum, which we cannot see. These different types of radiation have different wavelengths: the longest are gamma rays, and these are followed down the wavelength band by X-rays, ultraviolet light, visible light, infra-red rays, and radio waves, which have the shortest wavelength.

Only visible light and some radio waves can penetrate the Earth's atmosphere. An optical telescope is used to detect visible light, and a radio telescope, with its large dish-shaped reflector, is used to detect radio waves in space. The world's largest radio telescope is 1,000 ft in diameter and hangs in a hollow between hills at Arecibo, in Puerto Rico.

Telescopes in Space

The Earth's atmosphere can interfere with space observations. It blocks out light and filters out other forms of radiation that are useful to astronomers. To overcome this problem, instruments studying gamma rays, X-rays, ultraviolet rays, and infra-red rays, and some radio wave frequencies, are sent above the atmosphere, on balloons, sounding rockets, and satellites, to find out more about these sources of radiation in the Universe. A huge optical telescope, called the Hubble Space Telescope, was sent into orbit by the Space Shuttle in April 1990. It has a 94-in mirror, 300 mi above the Earth, and hopefully astronomers will see 14,000 million light years into the Universe.

Below: The Hale reflector telescope at Palomar, California, is over 200 in across, making it one of the largest in the world.

Above: The US Gamma Ray Observatory will study one of the many kinds of radiation in space.

LARGEST REFLECTING TELESCOPE HAS 20 FT MIRROR, IS SITED IN CAUCASUS MTS, USSR

Below: Radio waves are another form of radiation, and these are detected by huge dish aerials like this one at Jodrell Bank, England.

LARGEST RADIO DISH IS 1,000 FT IN DIAMETER, IS SITED AT ARECIBO, PUERTO RICO

THE NIGHT SKY

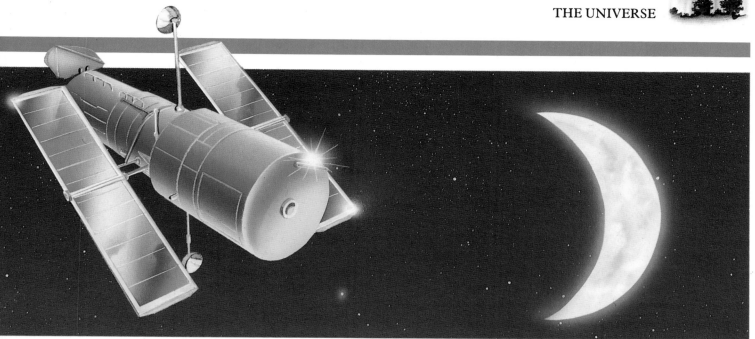

Left: High on a mountain, a large telescope scans the night sky. Above the Earth's atmosphere, a telescope in space sees the Universe in far greater detail.

Armed with even an ordinary pair of binoculars, there is a lot to see in the night sky. As well as the stars, you may also see the Moon, planets, shooting stars or meteors, star clusters, nebulas, galaxies, and, if you are very lucky, a comet.

You will notice that the stars vary in brightness and they seem to twinkle. This is caused by air currents which make the starlight dance around as it passes through the Earth's atmosphere. Stars also appear to have different colors. Type O stars are very white and these are the hottest stars (they have an estimated surface temperature of about 60,000 °F); then come cooler, type B bluish stars; type A, whitish stars; type F and G, yellow and very yellow stars, such as our Sun; type K, orange stars, and the coolest; type M, orange-red stars, which can have a surface temperature of 5,000 °F. The color depends mainly on the temperature of the star and partly on its chemical composition.

How to Look at the Night Sky

With naked eye observation, the best way to see something in the sky is to look slightly to the side of where the object is, rather than looking directly at it.

The Moon, especially when it is full, can interfere with star watching, but it is a great subject to look at itself, particularly when it is not full. During the phases of the Moon, an observer can easily see the craters, mountains, and other features on the lunar surface.

The planets Venus, Mars, Jupiter, and Saturn are often very prominent in the evening or morning sky, and at times it is even possible to spot Mercury. It is interesting to notice how Mars and the outer planets, particularly, appear to move in the sky, day by day, as they orbit the Sun.

Shooting stars are small pieces of rock called meteorites which enter the Earth's atmosphere at very high speed and burn up as friction is created against the air. You can see these streaks of light and, if you know where to look, faint patches of light, such as star clusters and nebulas. Comets can be seen occasionally, but many people, used to seeing photographs of them in books, find the real view rather disappointing.

Large manmade satellites, orbiting the Earth, can also be seen in the sky when they reflect the Sun's light. The ones that appear to flash are often the rocket stages that launched the satellites, tumbling out of control and catching the Sun's light as they rotate.

NEAREST STAR VISIBLE TO THE NAKED EYE IS ALPHA CENTAURI

CONSTELLATIONS

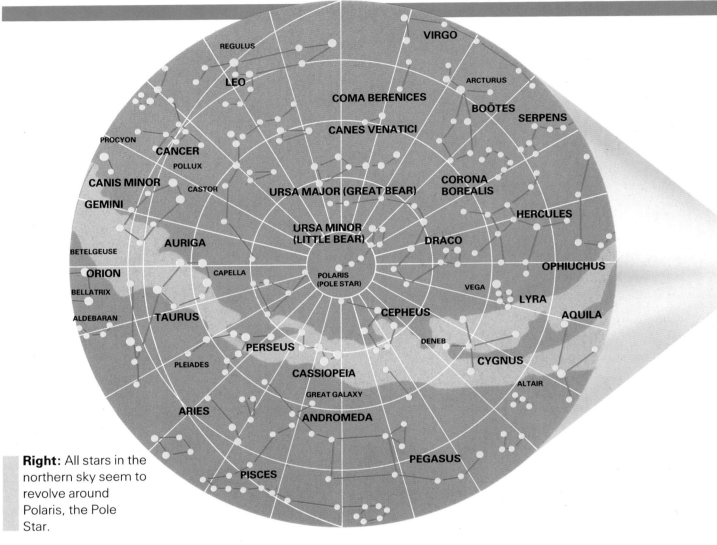

Right: All stars in the northern sky seem to revolve around Polaris, the Pole Star.

Many of the bright stars make distinctive patterns in the sky. These constellations are based mainly on star groups devised by Babylonian and Greek astronomers. They imagined that many of the star groups looked like characters from their mythology, and gave them suitable names such as Orion and the Great Bear. The Greek astronomer Ptolemy made a list of over 40 constellations in AD 150. The best-known constellations are those in the Zodiac, which is the area of sky that the Sun crosses. These include the easily identifiable constellations, such as Taurus the bull, as well as ones that are not so easy to make out, such as Aries the ram.

In the seventeenth and nineteenth centuries, new constellations were devised for the less defined star groups. In 1930, the International Astronomical Union issued the modern list of 88 constellations, from Andromeda to Vulpecula, which all astronomers now use.

Signposts in the Sky

The best way to start observing the stars is to use well-known constellations as signposts in the sky. Take the Great Bear, which is in the northern hemisphere. This includes the stars of the famous Big Dipper, which resembles a pot with a bent handle. If you look at the righthand part of the "pot" and draw an imaginary line upward across the two stars, Merak and Dubhe, the line points towards Polaris, the Pole Star, to which

CRUX – THE SOUTHERN CROSS – IS THE SMALLEST CONSTELLATION IN THE NIGHT SKY

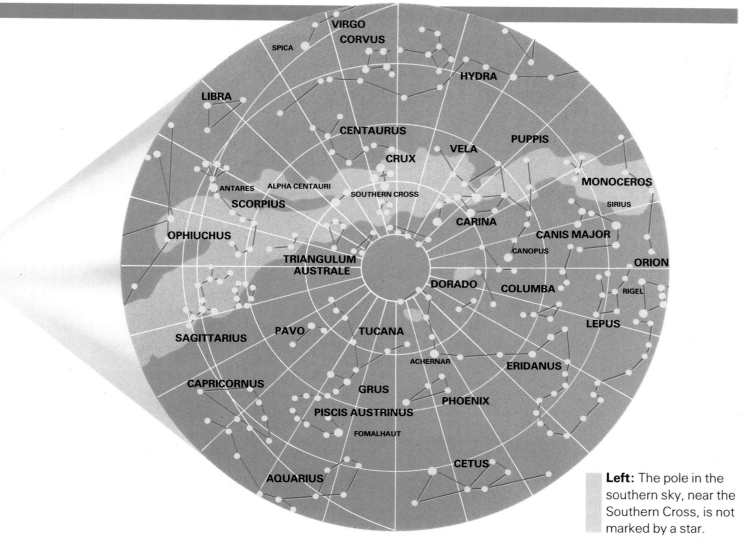

Left: The pole in the southern sky, near the Southern Cross, is not marked by a star.

the Earth's axis points. Polaris stays in the same position and all the other stars move around it.

If you follow an imaginary line from the two stars at the top of the "pot", Megrez and Dubhe, it points to the bright yellow star Capella in the constellation of Auriga. Follow an imaginary line leading upward from the stars at the left hand side of the "pot", Phekda and Megrez, and this leads to the bright white-blue star Vega in the constellation of Lyra. If you follow the bent part of the pot handle, an imaginary line points to the bright orange star Arcturus, in the constellation of Boötes.

In the southern hemisphere, the stars Alpha and Beta Centauri in Centaurus point to Crux, the South-

ern Cross, and the long axis of the cross points to the celestial south pole.

Probably the most magnificent constellation in the sky is Orion, which dominates the winter sky in the northern hemisphere. Look at this group of stars long enough and you cannot fail to imagine the hunter with his belt and sword, shield in hand, fighting off the bull Taurus nearby! Orion's two brightest stars are Rigel and Betelgeuse. His belt is marked by a line of three stars and below his belt hangs his sword – the Orion nebula. Taurus represents the head of a bull, his eye marked by the star Aldebaran. Taurus contains the two open clusters: the Pleiades and the Hyades.

HYDRA – THE WATER SNAKE – IS THE LARGEST CONSTELLATION IN THE NIGHT SKY

THE LIFE OF A STAR

Stars are glowing balls of gas that give out heat and light. They are formed from giant clouds of gas and dust called nebulas. The most famous, and the only one to be seen clearly with the naked eye, is the Orion nebula. It is situated about 1,600 light years away. In long-exposure photographs it is a spectacular, multi-colored object, featuring a black cloud like a horse's head, but with the naked eye it looks like a fuzzy patch in the constellation of Orion. Stars are forming today inside the Orion nebula, and radiation from the largest and hottest of these stars makes the nebula glow.

Star Birth
Stars are born when part of a gas cloud begins to break up. A fragment of gas shrinks under the inward pull of its own gravity. As it shrinks, becoming smaller and denser, pressures and temperatures build up inside until there is a nuclear reaction at the core in which hydrogen is turned into helium. This releases the energy that turns the ball of gas into a star. These nuclear reactions keep the star burning for the rest of its life.

It seems that stars are not born on their own but in groups. The most famous star cluster is the Pleiades in the constellation of Taurus. It consists of over 200 stars, although only seven can be seen with the naked eye, so it is sometimes called the Seven Sisters.

Also in our Galaxy are much larger, ball-shaped clusters of very old stars. These globular clusters include Praesepe or the Beehive cluster in the constellation of Cancer, and the M13 globular cluster in the constellation of Hercules. The brightest globular clusters can only be seen in the southern hemisphere. They include Omega Centauri and 47 Tucanae. With the naked eye, globular clusters look like misty patches, but they contain many hundreds of thousands of stars. These are some of the oldest known stars, which formed more than 10,000 million years ago.

Star Death
The lifetime of a star depends on its mass. Heavy stars burn out more quickly than light ones because they burn their hydrogen "fuel" more quickly. When a star begins to run out of hydrogen at its core, the nuclear reactions move outward. The extra energy makes the star expand so the outer layers become cooler and redder until the star becomes a huge red giant. Gradually, the outer layers of this enormous star drift off into

Left: The beautiful Pleiades star cluster, sometimes called the Seven Sisters. These young stars, only 60 million years old, are still surrounded by the cloud of dusty gas from which they formed.

LARGEST STAR IS RED SUPERGIANT BETELGEUSE, 400 MILLION MI IN DIAMETER

Left: The Trifid nebula in the constellation of Sagittarius is a birthplace of stars.

Below: The life of a star like our Sun, which begins as a ball of gas and dust and ends as a dwarf.

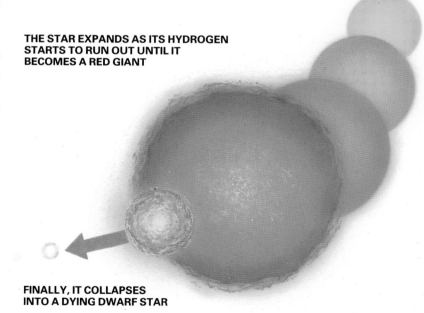

A STAR FORMS
OUT OF INTERSTELLAR
GAS AND DUST

IT CONTRACTS AS IT BEGINS
TO WORK LIKE A STAR,
PERHAPS FOR 30 MILLION YEARS

THE STAR EXPANDS AS ITS HYDROGEN
STARTS TO RUN OUT UNTIL IT
BECOMES A RED GIANT

FINALLY, IT COLLAPSES
INTO A DYING DWARF STAR

space, forming a giant smoke ring. Stripped of the outer layers, the small, hot core of the star can be seen. This is called a white dwarf star. It is extremely dense; a spoonful would weigh several tons. Dwarf stars are about one-hundredth the diameter of our Sun and include the brightest star in the sky, Sirius.

A star more massive than the Sun turns into a red supergiant; these are larger and brighter than the red giants. They include Antares and Betelgeuse which are about 500 times larger than our Sun. Although they have a cool surface temperature, they are very bright because of their size. A series of runaway nuclear reactions at the core of a red supergiant results in a gigantic explosion called a supernova. In 1054, astronomers in China saw just such an explosion in the constellation of Taurus. It shone so brightly that it was visible in daylight for three weeks. The shattered remains of that star can be seen in telescopes as a fuzzy patch known as the Crab nebula.

A supernova does not always completely destroy a star – sometimes the core remains as a tiny, compressed object called a neutron star. This is even smaller and denser than a dwarf star.

SMALLEST STAR IS WHITE DWARF STAR L362–81, 3,500 MI IN DIAMETER

COSMIC MYSTERIES

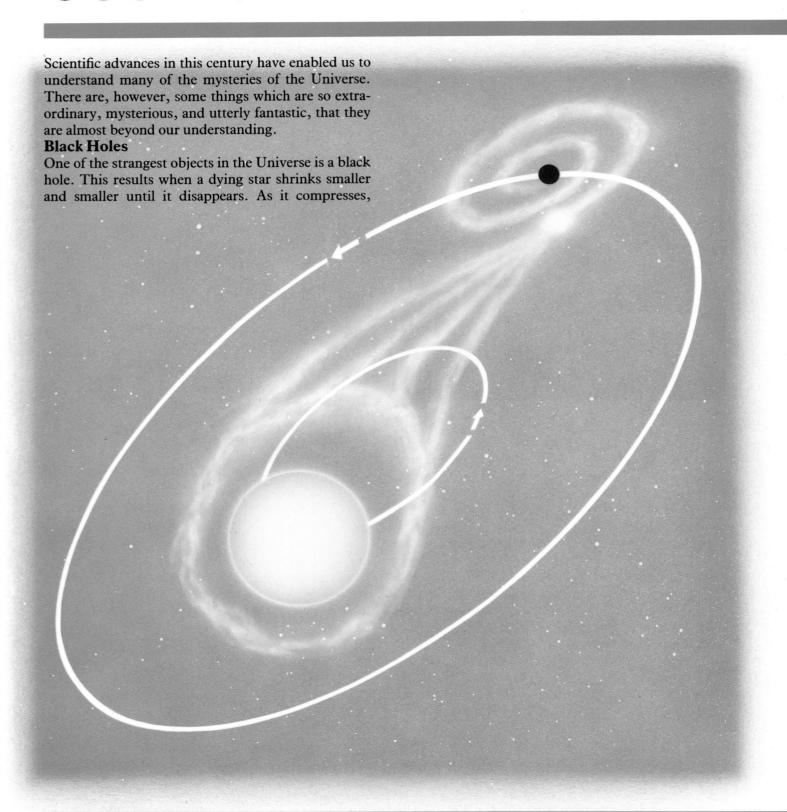

Scientific advances in this century have enabled us to understand many of the mysteries of the Universe. There are, however, some things which are so extraordinary, mysterious, and utterly fantastic, that they are almost beyond our understanding.

Black Holes

One of the strangest objects in the Universe is a black hole. This results when a dying star shrinks smaller and smaller until it disappears. As it compresses,

CYGNUS 1 – FIRST BLACK HOLE TO BE DISCOVERED, ORBITS A BLUE STAR HDE 226868

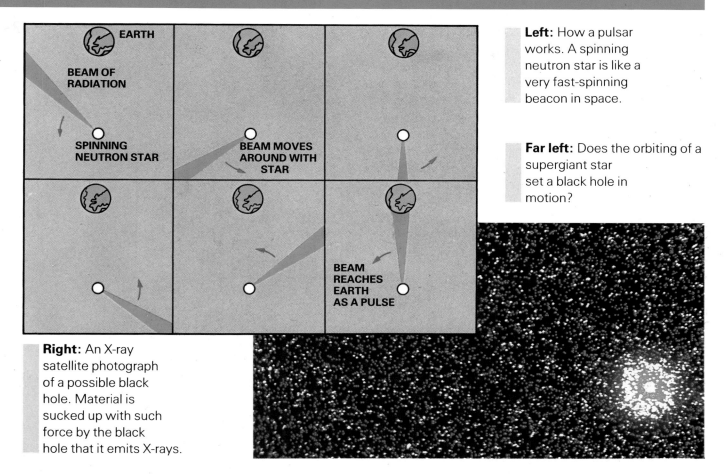

EARTH

BEAM OF RADIATION

SPINNING NEUTRON STAR

BEAM MOVES AROUND WITH STAR

BEAM REACHES EARTH AS A PULSE

Left: How a pulsar works. A spinning neutron star is like a very fast-spinning beacon in space.

Far left: Does the orbiting of a supergiant star set a black hole in motion?

Right: An X-ray satellite photograph of a possible black hole. Material is sucked up with such force by the black hole that it emits X-rays.

space and time are sucked in by an enormous gravitational force. Once inside, nothing can get out again – not even light can escape. The star has shrunk itself out of existence. All that remains is the star's intense gravitational force, forming a black hole that sucks up material like a cosmic vacuum cleaner. Inside, space and time might become so distorted that they actually cease to exist. So how can black holes be detected? They suck up material with such force that it is heated up millions of degrees, and becomes so hot that it emits X-rays. These can be detected by satellites orbiting the Earth.

Neutron Stars and Pulsars

Some smaller stars which collapse do not turn into black holes but into neutron stars. These are very small, very dense stars, which are as heavy as our Sun but only a few miles in diameter; a spoonful of neutron star material would weigh a thousand million tons. Neutron stars can spin very quickly – once a second or less – and many of them send out radio waves like the flash of light from a lighthouse. These radio waves are received by radio telescopes on the Earth. Such flashing neutron stars are called pulsars. One of the most famous pulsars is at the center of the Crab nebula, in the constellation of Taurus. The Crab Pulsar flashes at the rate of 30 times per second. It is estimated to be about 1,000 years old and was born in a supernova – the explosion of a star.

Quasars

The quasar is one of the most puzzling phenomena in the Universe. It is a tiny point of light that sends out more energy than a whole galaxy. Some astronomers think that they may be young galaxies with an immense black hole in the center which rips apart passing stars and swallows their gas. The furthest quasars are thought to be the most distant objects visible in the Universe.

QUASARS, REMOTEST OBJECTS FOUND IN UNIVERSE, 14,000 MILLION LIGHT-YEARS AWAY

THE SOLAR SYSTEM

Our Sun is a small star at the center of the Solar System. This consists of nine known planets, moons, asteroids, comets, and meteorites, all of which orbit the Sun. They are kept in the Solar System by the Sun's immense gravitational pull. Astronomers believe that the Solar System was formed about 4,500 million years ago.

Formation of the Solar System

There was once a huge cloud of gas and dust containing mainly hydrogen, some helium, and other elements. The cloud began to spin and, as it did so, it began to break up and finally collapsed inward under the pull of its own gravity. In the center was a core surrounded by a disk of dust and gas. Specks of dust

Right: The nine known planets in the Solar System: 1 Mercury, 2 Venus 3 Earth, 4 Mars, 5 Jupiter, 6 Saturn, 7 Uranus, 8 Neptune, 9 Pluto. Jupiter is the largest, Pluto is the smallest and coldest, Venus is the hottest.
Jupiter, Saturn, Uranus, and Neptune have ring systems, the most prominent being Saturn's.

99.86 PER CENT OF THE MASS OF THE SOLAR SYSTEM RESIDES IN THE SUN

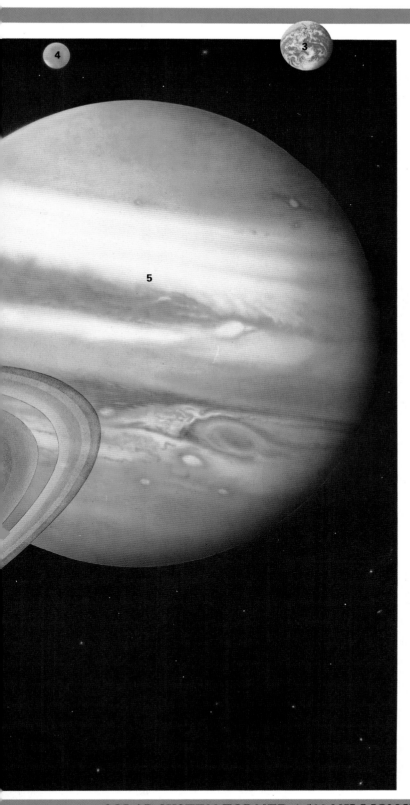

began to join together to form tiny rocks, first by colliding with each other, and then by the gravitational pull of other rocks. This process continued for millions of years until the planets were formed with gas from the cloud making up their atmospheres. The central core, meanwhile, had become dense and hot, and an explosion turned it into a star. The Solar System was formed with the planets orbiting the Sun.

The Solar System

Going out from the Sun, the planets are: Mercury, Venus, Earth, Mars, Jupiter, Saturn, Uranus, Neptune, and Pluto. Mercury, Venus, Earth, and Mars are the rocky planets. Jupiter, Saturn, Uranus and Neptune are called the giant planets. These have huge atmospheres with tiny rocky cores. These planets were just large enough and had just the right mix of gases to have almost become stars themselves. Some astronomers believe that Jupiter almost became a star which would have made the Sun a "Double Star". If Jupiter had become a small sister of the Sun, then the planets relatively close by would not exist. Earth would have turned out vastly different too! Pluto is unusual because it orbits the Sun at a greater distance than Neptune, but is not a large planet with a huge atmosphere. Some astronomers believe that it used to be a moon of Neptune. There are unlikely to be any undiscovered giant planets in the outer Solar System beyond Pluto, but there could possibly be small rocky planets, like Pluto, yet to be discovered.

Between Mars and Jupiter there is a band of large rocks known as asteroids. Some of these appear to have been captured by the gravity of some of the planets and have become moons. There are also smaller pieces of rocks which are meteorites, while other leftover material became balls of gas and dust, which, when they come close to the Sun, are heated up, losing material and forming tails. These are the comets.

The End of the Solar System

Scientists predict that the Sun may last another 5,000 million years. Then it will run out of hydrogen fuel and expand into a red giant, enveloping many of the planets, such as the roasted Earth. The Sun will become denser and explode, destroying all of the Solar System, leaving a tiny white dwarf star, which will gradually cool and fade until it becomes a cold cinder.

SOLAR SYSTEM FORMED 4,500 MILLION YEARS AGO FROM A GAS AND DUST CLOUD

THE SUN

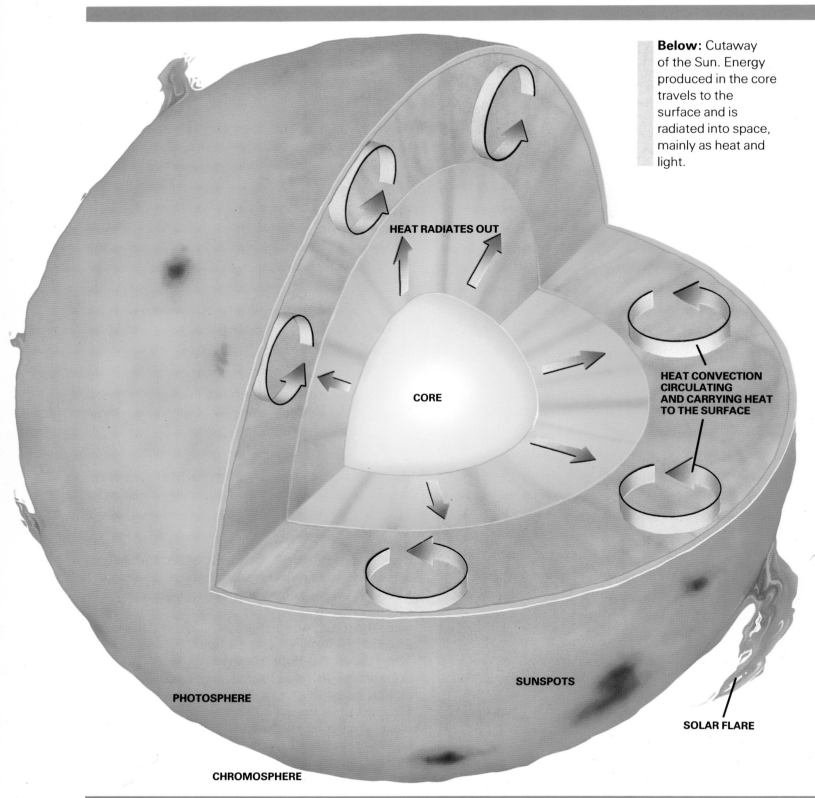

Below: Cutaway of the Sun. Energy produced in the core travels to the surface and is radiated into space, mainly as heat and light.

HEAT RADIATES OUT

CORE

HEAT CONVECTION CIRCULATING AND CARRYING HEAT TO THE SURFACE

SUNSPOTS

SOLAR FLARE

PHOTOSPHERE

CHROMOSPHERE

93 MILLION MI AWAY • DIAMETER 870,000 MI • TURNS ON ITS AXIS IN 25.4 DAYS

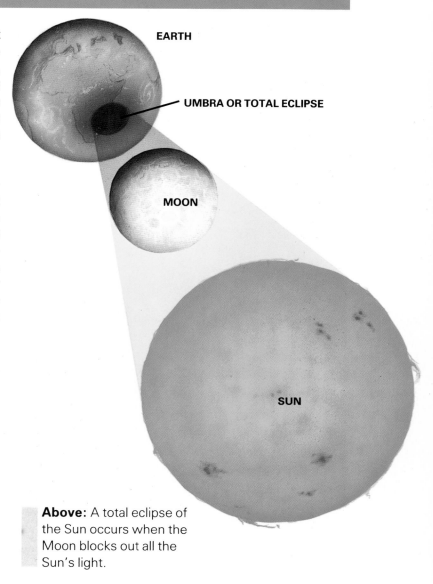

At night, thousands of stars can be seen in the sky, but during the day only one star is visible: our Sun. Without its heat and light, the Earth would be dark, cold, and lifeless. The Sun is an average-size star but it looks bigger and brighter than the night-time stars because it is so much closer. Light takes 8 minutes 17 seconds to reach the Earth from the Sun, but more than 4 years from the next nearest star, Proxima Centauri.

The Giant Nuclear Furnace

Planets are cold, dark bodies that only shine by reflecting sunlight; stars produce their own light and heat through nuclear reactions. At the center of the Sun is a giant nuclear furnace. Here atoms of hydrogen gas combine or fuse together to form helium. This process, called nuclear fusion, releases enormous amounts of energy as light and heat. It is this energy which makes the Sun shine. The temperature at the center of the Sun rises to many millions of degrees, but at the surface is only about 11,000 °F.

The visible surface of the Sun, called the photosphere, is not solid, but a seething mass of hydrogen gas. It is often marked by dark patches known as sunspots. These are the sites of violent storms and eruptions which become more frequent every 11 years or so. Storms also cause flames of hot gas, called prominences, to leap thousands of miles into space from the surface of the Sun.

The Sun is enveloped in a halo of gas called the corona, which is seen glowing around the Sun at total eclipses. It consists of hot gas boiled off from the Sun's surface. Gas from the corona streams away from the Sun forming the solar wind that blows past Earth and other planets.

Auroras

When sunspots occur, the solar wind blows stronger than usual, so streams of atomic particles reach Earth. They collide with atoms in the upper air causing them to give off colored light: this magnificent spectacle is called an aurora. It looks like a curtain, usually red or green in color, which stretches for thousands of miles, often with folds that seem to move and shimmer. Usually, auroras only appear near the Arctic and Antarctic Circles. In the northern hemisphere, it is called the Aurora Borealis, or northern lights; in the southern hemisphere, it is called the Aurora Australis,

Above: A total eclipse of the Sun occurs when the Moon blocks out all the Sun's light.

or southern lights.

Solar Eclipse

When the Moon passes in front of the Sun, cutting off some or all of the Sun's light, an eclipse of the Sun takes place. Sometimes the Moon completely blocks out the Sun's light, making day suddenly turn to night; this is a total eclipse.

Warning

It is dangerous to look at the Sun with the naked eye. NEVER look through binoculars or a telescope at the Sun or you will immediately be blinded.

CENTER OF SUN IS 30 MILLION °F • BURNS 4 MILLION T OF HYDROGEN PER SECOND

MERCURY

Imagine standing on a desolate, cratered world, where the day lasts 176 Earth days, the Sun appears to wander around the sky and the temperature is 750 °F. This is Mercury, the closest planet to the Sun. It would be impossible to walk on the planet.

Mercury is a small, rocky body measuring 2,950 mi in diameter. This is smaller than the Earth and only 50 per cent larger than the Moon. However, it is as heavy as the Earth. Mercury rotates on its axis once every 58.65 Earth days and takes 87.97 Earth days to orbit the Sun. This results in the Mercury day lasting 176 Earth days. Night lasts the same length of time and, because there are no clouds to keep the heat in, temperatures drop to −290 °F.

Mercury travels as close to the Sun as 28.5 million mi and as far as 43.3 million mi. These near and far values are known as perihelion and aphelion. The Sun will be almost twice the size in the sky when Mercury is at perihelion, compared with its aphelion. Although Mercury can appear bright in the Earth's sky, it is difficult to see because it is so close to the Sun.

Mariner 10 Visits Mercury

Astronomers only found out what the planet looked like in 1974. The American spaceprobe Mariner 10 became the first and, so far, only Mercury explorer. Mariner 10 was also the first probe to visit two planets in turn – Venus and Mercury. It used the gravitational pull of Venus to swing it into a different orbit. Its flight path took it past Mercury three times in 1974–75. Mariner's 2,800 photographs revealed that the surface of Mercury looks rather like the Moon. It is covered with craters but does not have as many large, smooth areas. Photographs also showed the Caloris Basin. This is a huge 800 mi diameter crater, surrounded by 6,560 ft high mountains. Mercury probably has a large core that forms about 80 per cent of its body. This core consists of nickel and iron.

Mercury was once a cooling, molten globe, and, like other planets and the Moon, was bombarded with rocks during the violent formation of the Solar System. These blasted-out craters on the surface of the planet. Mercury is so close to the Sun that it is airless and waterless; this means there has been no erosion to destroy these craters. On the Earth, evidence of this cratering has been eroded by the atmosphere.

Right: Mariner 10 discovered that Mercury was covered with craters.

Top: This photograph has been tinted to show what Mercury would look like to the human eye.

Below: Mariner 10 was the only spacecraft to re-visit a planet. It made three fly-bys of Mercury in 1974–75. No other spacecraft has been back – yet.

CIRCLES SUN IN 87.97 DAYS • DIAMETER 2,950 MI • TURNS ON ITS AXIS IN 58.65 DAYS

VENUS

The planet Venus is the brightest object in the Earth's sky after the Sun and the Moon; it can be bright enough to be seen during the day. As it is almost the same size as the Earth, it used to be regarded as our sister planet. Venus is veiled in a mysterious, white blanket of cloud. Before spacecraft visited Venus, the planet was often regarded as a beautiful one, possibly with rich, lush vegetation and a hot climate. Venus is actually a hell-hole!

The planet has an almost perfectly circular orbit around the Sun, at a distance of 67.2 million mi. It takes 224.7 days to travel once around the Sun, which is the Venus year. It rotates only once every 243 Earth days and, because of the angle at which it rotates, a day on Venus lasts 116 Earth days. Venus is unusual because it rotates from east to west, which is the opposite direction to the Earth and other planets. It can come to within 25 million mi of Earth, closer than any other planet. This and its reflective white clouds make it appear exceptionally bright.

The dense atmosphere surrounding Venus consists of unbreathable carbon dioxide gas. This results in the "greenhouse effect": the atmosphere lets in the Sun's heat but does not let it escape so the surface temperature on Venus is a furnace-like 880 °F! The pressure of its atmosphere on the surface is about 90 times that of the Earth's atmospheric pressure. The serene-looking blanket of cloud seen through telescopes is in places a violent gale. The clouds consist of strong sulphuric acid, and horrendous lightning strikes the surface. Venus is one of the most inhospitable places imaginable. An astronaut who crashed on Venus would be instantly crushed, roasted, and suffocated, and the remains of his or her body would be eaten away by a rain of acid droplets.

Spaceprobes Visit Venus

The first spacecrafts which attempted to land were crushed by the atmospheric pressure. Yet finally several Soviet spaceprobes have managed to land on the surface and transmit photographs. These have shown clear views of eroded, flattened boulders and rocks. Other spacecrafts have orbited the planet equipped with radar mapping cameras. Their radar beams penetrated the thick cloud layer to produce a map of the planet.

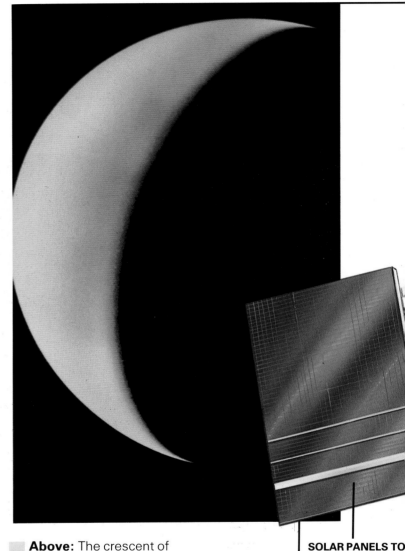

Above: The crescent of Venus, reflecting sunlight off its thick, white blanket of cloud, which makes it appear exceptionally bright.

SOLAR PANELS TO PROVIDE ELECTRIC

Over half the surface consists of rolling plains of boulders and rocks, but there is also a huge highland region in the northern hemisphere called Ishtar Terra, which measures 1,800 mi across. This includes a mountain, Maxwell Montes, which is 7.5 mi high – higher than Mount Everest on Earth.

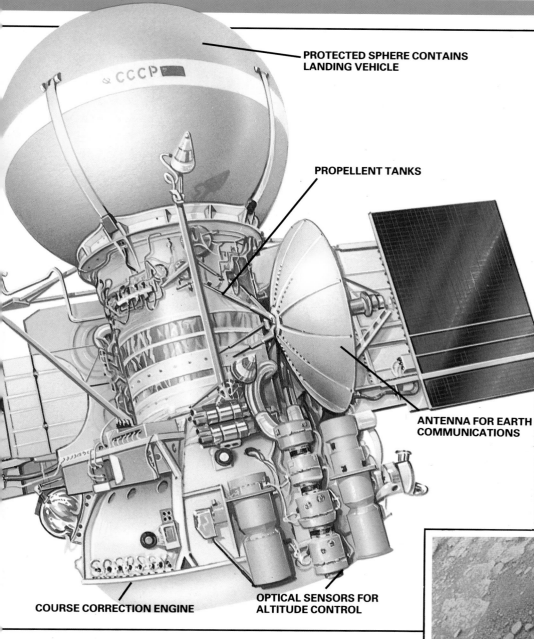

PROTECTED SPHERE CONTAINS LANDING VEHICLE

PROPELLENT TANKS

ANTENNA FOR EARTH COMMUNICATIONS

COURSE CORRECTION ENGINE

OPTICAL SENSORS FOR ALTITUDE CONTROL

VENUS PROBES

Venus was the first target for probes because of its closeness to Earth. The first successful space-probe to another planet was the American Mariner 2, launched toward Venus in August 1962. It flew past Venus at a distance of 21,000 mi on 14 December 1962.

The USSR has specialized in the space-probe exploration of Venus. Venera 4, 5, and 6 were crushed by the intense pressure when they reached the surface of the planet. Venera 7 made the first successful landing in 1970. Cameras were carried to the surface of Venus for the first time by Venera 9 and 10 in October 1975.

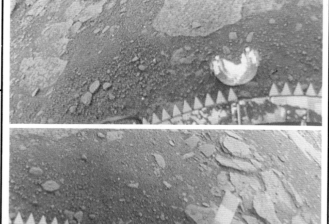

Above: A typical Soviet Venera spacecraft. It entered orbit around Venus, and the lander was jettisoned from the spherical section on top of the spacecraft.

Right: The surface of the planet Venus taken by the Soviet Venera 13 landing craft in March 1982. Part of the spacecraft is visible in the foreground.

CIRCLES SUN IN 224.7 DAYS • DIAMETER 7,518 MI • TURNS ON ITS AXIS IN 243 DAYS

EARTH AND MOON

The third planet from the Sun is a bright blue, star-like object called the Earth, which has its own less bright companion moon. Three-quarters of the Earth is covered with water, and the land appears green in parts but very barren in others. There are mountain ranges, some with caps of white ice. The Earth's atmosphere contains fast-moving white clouds made of water vapor. Its oxygen-nitrogen atmosphere allows sunlight and heat to reach the surface, but a layer of ozone cuts out some of the harmful radiation from the Sun.

Earth is unique among the planets as it is the only one on which life exists. It is the home of 5,000 million human inhabitants and an incredible variety of plants and animals. The reason why there is life on Earth, is because it is at just the right distance from the Sun for water to exist in liquid form.

The Moon

Earth's companion satellite is a gray, cratered, airless and waterless ball of rock. It is 2,160 mi in diameter and lies 238,855 mi away. It orbits the Earth about every 28 days, which is called a month. The Moon is important because its gravity causes tides on the Earth and provides light at night. It rotates at the same speed as it orbits the Earth, so that only one face of the Moon can be seen from Earth. The Moon displays different phases through the month, from a fine crescent to a bright, fully illuminated disk, depending on the position of the Moon in relation to the Sun. This is because the Moon does not give off light itself, but only shines by reflecting sunlight. From Earth, the Moon is almost the same size in the sky as the Sun. This results in some spectacular eclipses when the Moon passes in front of the Sun.

To the naked eye, the Moon appears dotted with dark patches. These are lowland plains, known as the Moon's seas, and are filled not with water but with lava that oozed out from inside the Moon thousands of millions of years ago. The smooth, dry seas are given Latin names, such as Mare Tranquillitatis (Sea of Tranquillity) or Oceanus Procellarum (Ocean of Storms). The bright areas of the Moon are rugged mountains and craters. The largest craters are 60 mi or more in diameter and are named after famous scientists of the past. It is thought they were formed by meteorites that crashed into the Moon. The Moon's surface has not changed for the past 3,000 million years.

PLANET EARTH

One year: 365 days.
Diameter: 7,926 mi.
Crust: 20 mi thick.
Core: 3,000 mi deep, molten iron.
Atmosphere: 78 per cent nitrogen, 21 per cent oxygen.
Deepest point: The Mariana Trench, 35,830 ft.
Highest point: Mount Everest, 29,021 ft.
Speed of rotation: 1,030 mi/h.
Weight: 6,095 million million million tons.

Right: Craters on the far side of the Moon taken during an Apollo mission. When the astronauts were on the far side, they were totally cut off from communications with the Earth.

THIRD PLANET FROM SUN • AVERAGE DISTANCE FROM SUN 93 MILLION MI

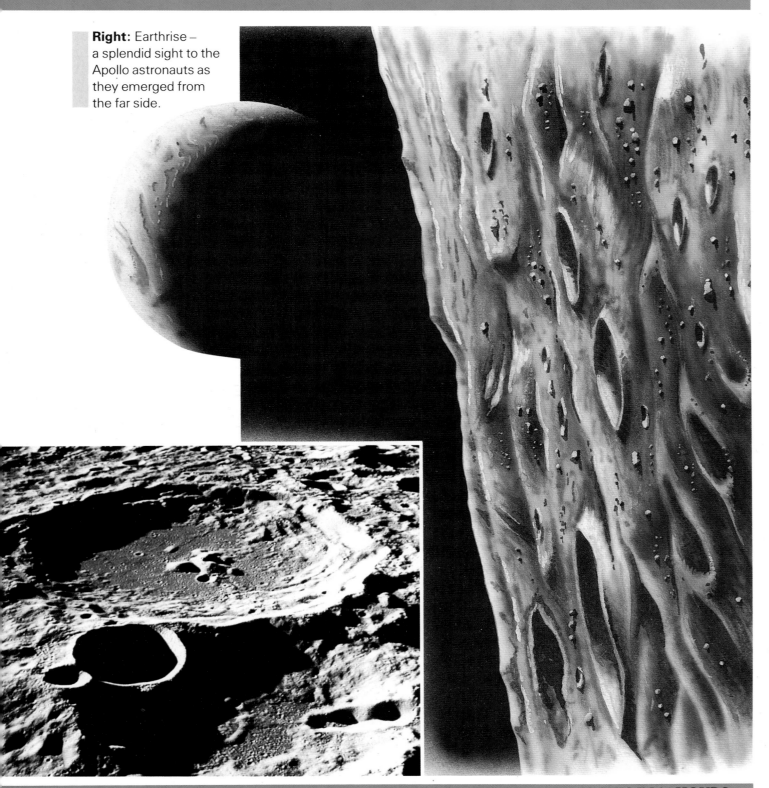

Right: Earthrise – a splendid sight to the Apollo astronauts as they emerged from the far side.

CIRCLES SUN IN 365.25 DAYS • DIAMETER 7,926 MI • TURNS ON ITS AXIS IN 24 HOURS

33

MARS

From the Earth, the planet Mars looks like a red star in the sky, so it is also known as the red planet. Because so much of its surface can be observed even with a relatively small telescope, this planet has captured the imagination of many science fiction writers. Earlier this century, some astronomers reported seeing a network of canals on Mars, and dark patches on the surface which they thought were patches of vegetation. They speculated that a Martian civilization built the canals to bring water from the poles to irrigate the deserts of Mars so they could grow crops. Soon, the idea of not just life, but intelligent life on Mars was promoted. Now, from the results of the landing on Mars, in 1976, of the US Viking 1 and 2 spacecraft, we know that life does not exist there. The dark areas are simply areas of darker rock and dust, and there is no sign of the canals.

Mars is 4,200 mi in diameter and orbits the Sun every 687 days – a Martian year. Mars has a similar length of day to the Earth: 24 hours 37 minutes 23 seconds. Climatically, it is very different. Its very thin atmosphere is 95 per cent carbon dioxide, and has an atmospheric pressure less than 1 per cent that of Earth's. Temperatures vary from −20°F to −202°F. It has very active weather, with winds churning up the

Below: This photograph of Mars, was taken by Viking 2 in 1976. On the left, with frozen cloud plumes, is the giant volcano Ascraeus Mons. In the center is the great rift canyon Mariner Valley (Vallis Marineris).

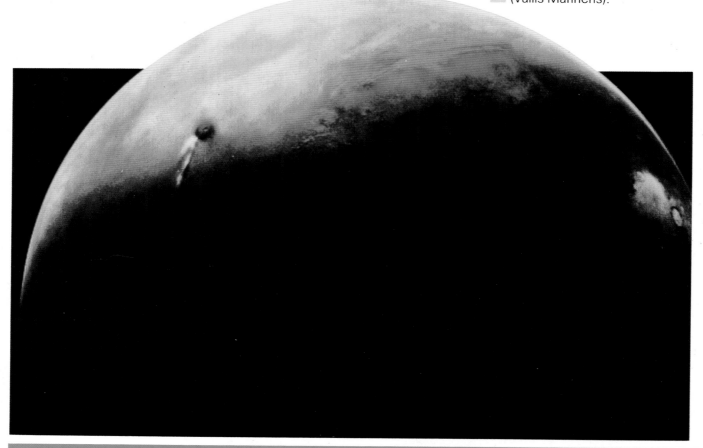

FOURTH PLANET FROM THE SUN • AVERAGE DISTANCE FROM SUN 141.5 MILLION MI

VIKING SPACEPROBES

In 1975, two identical Viking spaceprobes were launched. They consisted of an orbiter to survey the planet from above, and a lander to perform experiments on the ground. The first Viking lander touched down on July 20 1976 in a lowland area called Chryse (right) over which liquid water was believed to have flowed in the past when Mars was wetter. Vikings 1 and 2 took color photographs of the landscape, instruments aboard measured wind speed and air temperatures, and they tested soil samples to detect life; none was found.

red soil and turning the sky pinkish. The red color of the rocks and soil results from their high content of iron oxide. Mars' ice polar caps are a mixture of frozen carbon dioxide and ice.

Spaceprobes Visit Mars

The first successful spacecraft to visit Mars was Mariner 4 in 1965. It returned photographs which showed that Mars was heavily cratered. Mariner 9 orbited the planet in 1971 (it was the first probe to orbit another planet) and mapped the surface. In 1976, two Viking spacecraft landed on Mars and revealed an extraordinary array of features, from huge canyons to high volcanoes. The largest of the volcanoes, called Olympus Mons, is 16 mi high and is the largest volcano in the Solar System. One of the canyons, called Mariner Valley, is 3.4 mi deep, up to 75 mi wide and 2,500 mi long.

The Moons of Mars

Mars has two moons which appear to be asteroids captured by its gravity, as they are potato-shaped, cratered lumps of rock, rather than spheres. The moons are called Deimos and Phobos. Phobos measures 17 mi by 13 mi by 12 mi. Deimos measures 9 mi by 7 mi by 6 mi.

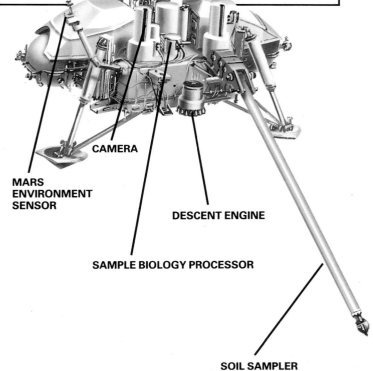

CAMERA

MARS ENVIRONMENT SENSOR

DESCENT ENGINE

SAMPLE BIOLOGY PROCESSOR

SOIL SAMPLER

Above: A Viking lander

CIRCLES SUN IN 687 DAYS • DIAMETER 4,217 MI • TURNS ON ITS AXIS IN 24.62 HOURS

ASTEROIDS

When the planets in the Solar System were formed, large pieces of rock were left orbiting the Sun. Many collided with each other, forming smaller fragments. These are called asteroids. They are rocky bodies, irregularly shaped, and pitted with craters. There could be more than 50,000 asteroids – many may be too small and too faint to be seen. About 2,500 of these form a belt which orbits the Sun between the planets Mars and Jupiter.

The Largest Asteroids

The largest asteroid is called Ceres. It was discovered by the Italian astronomer, Giuseppe Piazzi in 1801. It is about 582 mi at its widest point and orbits the Sun every 4.6 Earth years. The next asteroid to be discovered, a year later, is called Pallas. It is about 330 mi at its widest point. The brightest asteroid is called Vesta and is just visible with the naked eye. Vesta is almost the shape of a planet, is about 322 mi in diameter and takes 3.64 years to orbit the Sun. Astronomers have also discovered that Vesta rotates every 10 hours. The next largest asteroid is Hygeia, which was discovered in 1849. This is 254 mi at its widest point. The 224 mi wide Davida was discovered in 1903.

Other Asteroids

There are other groups of asteroids which have a highly elliptical orbit, traveling furthest away from the Sun in the asteroid belt but also coming very close to the Sun. One of these is called Icarus, after the mythological character who tried to fly close to the Sun and perished. This was discovered in 1949. It is only 0.86 mi at its widest point, and may be almost spherical. It rotates every 2.5 hours. During its journey around the Solar System, Icarus comes to within 18 million mi of the Sun, close enough to be heated to a glowing red. It orbits the Sun every 1.1 years and occasionally passes very close to the Earth. In 1968, Icarus passed to within 3.9 million mi of the Earth, and could just be seen, like a faint star, whizzing across the sky.

Another asteroid which travels close to the Earth is Eros. This cigar-shaped asteroid is about 22 mi long by 4 mi wide and rotates every five hours. It is quite possible that asteroids like Icarus and Eros have passed very close to the Earth, causing extraordinary climatic changes, or even hit the Earth itself, causing huge craters. The cause of the massive explosion, equivalent to a 12-megaton nuclear bomb, that hit Siberia in 1908, is still unexplained and could have been an asteroid.

Right: The asteroids that orbit the Sun between Mars and Jupiter. Four interplanetary spacecraft have flown through the asteroid belt without being damaged.

LARGEST ASTEROID IS CERES, DIAMETER 582 MI, 260 MILLION MI FROM SUN

JUPITER

The largest planet in the Solar System, Jupiter is a huge ball of mainly gaseous hydrogen and helium. It weighs 2.5 times as much as all the other planets rolled together. From the Earth, it appears as a bright yellow "star" in the night sky, brighter than the brightest real star, Sirius, but not nearly as bright as Venus.

Before the space age, all that we knew about the planet was that it orbited the Sun at an average distance of 483,365,000 mi, had 12 moons, had an extraordinary feature called the Great Red Spot among its multi-colored bands of clouds and rotated once in just under 10 hours.

This extremely fast rotation rate – the fastest of any planet – results in a flattening at the planet's poles.

Spaceprobes Visit Jupiter

Thanks to the explorations of Pioneers 10 and 11 and Voyagers 1 and 2 between 1973 and 1979, astronomers now know much more about the planet. Jupiter has 16 not 12 moons. It also has rings around it, made up of small particles held in place by the gravitational pull of the planet. The planet has an extremely turbulent upper atmosphere of swirling bands of clouds, some whizzing around the planet at a rate of 22,000 mi/h. These bands are red, brown, yellow and, sometimes, even blue from the chemicals, such as ammonium and hydrogen sulphide, in the atmosphere. Phosphorus is

JUPITER'S MOONS

The four largest moons of Jupiter – Io, Europa, Ganymede and Callisto – are sometimes referred to as the Galilean satellites because they were discovered by Galileo in 1610. They can be seen through a pair of binoculars. Io, 2,260 mi in diameter, has an extraordinary orange, yellow, and white surface with deposits of sulphur and sulphur dioxide, as well as volcanoes which spew sulphur dust and gas 170 mi into space. Europa, with a diameter of 1,900 mi, has a smooth surface covered completely with ice, parts of which are streaked with cracks. Ganymede, the largest of Jupiter's planets, with a diameter of 3,300 mi, has a surface of ice and rock, valleys 10 mi wide and mountains 3000 ft high. Callisto, 3,000 mi in diameter, is the most heavily cratered body yet observed of Jupiter's planets.

FIFTH PLANET FROM THE SUN • AVERAGE DISTANCE FROM SUN 483.6 MILLION MI

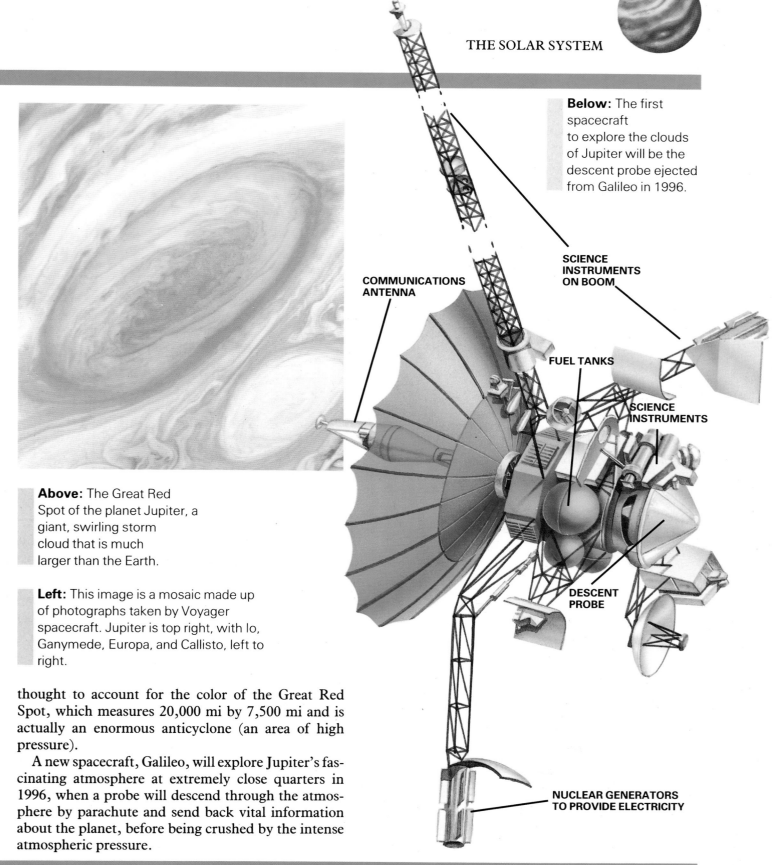

Below: The first spacecraft to explore the clouds of Jupiter will be the descent probe ejected from Galileo in 1996.

SCIENCE INSTRUMENTS ON BOOM

COMMUNICATIONS ANTENNA

FUEL TANKS

SCIENCE INSTRUMENTS

DESCENT PROBE

NUCLEAR GENERATORS TO PROVIDE ELECTRICITY

Above: The Great Red Spot of the planet Jupiter, a giant, swirling storm cloud that is much larger than the Earth.

Left: This image is a mosaic made up of photographs taken by Voyager spacecraft. Jupiter is top right, with Io, Ganymede, Europa, and Callisto, left to right.

thought to account for the color of the Great Red Spot, which measures 20,000 mi by 7,500 mi and is actually an enormous anticyclone (an area of high pressure).

A new spacecraft, Galileo, will explore Jupiter's fascinating atmosphere at extremely close quarters in 1996, when a probe will descend through the atmosphere by parachute and send back vital information about the planet, before being crushed by the intense atmospheric pressure.

CIRCLES SUN IN 11.9 YEARS • DIAMETER 88,730 MI • TURNS ON ITS AXIS IN 9.84 HOURS

SATURN

Saturn is probably the most spectacular planet of the Solar System. It was once known as the ringed planet, the only body in the Solar System with a band of orbiting particles. Through advances in astronomical observation and visits by spacecraft, we now know that Jupiter, Uranus, and Neptune also have ring systems, but they are far less impressive than those of Saturn.

Saturn is the second largest planet in the Solar System. It measures 75,100 mi in diameter and takes 29.5 years to make one journey around the Sun, at an average distance of 886 million mi. It rotates on its axis in just over 10 hours. Saturn is a huge ball of mainly hydrogen gas. Bands of clouds, like those around Jupiter but less distinct, swirl over the planet.

Beneath the clouds are thick lakes of liquid hydrogen. At the center of the planet is a rocky core, about 12,000 mi in diameter; this is larger than the Earth.

Saturn's Ring System

The ring system around the planet is about 170,000 mi wide but only 300 ft deep. Through a telescope it seems to be divided into five separate systems.

Right: From Earth, five large rings can be seen circling Saturn, but Pioneer 11 and Voyager I and 2 discovered many more rings, too faint to be seen from Earth.

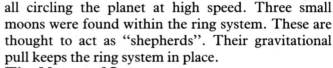

However, when the US Pioneer 11 spaceprobe passed Saturn in 1979 and the Voyager 1 and 2 spaceprobes flew by in 1980 and 1981, the ring system proved to be much more complicated. It consisted of thousands of narrow ringlets, each one made up of thousands of millions of "snowballs" of ice and rock. These range from very small flakes to particles about 32.8 ft wide, all circling the planet at high speed. Three small moons were found within the ring system. These are thought to act as "shepherds". Their gravitational pull keeps the ring system in place.

The Moons of Saturn

Before spaceprobes traveled to Saturn, it was thought to have had 10 moons but it is now known to have at least 17. The largest is Titan, which measures 3,000 mi in diameter. The surface is obscured by a dense, cloudy atmosphere that scientists thought was methane, but the Voyager spaceprobes revealed that it is mainly nitrogen, and measured temperatures of -290 °F on the surface. Photographs from Voyager showed that the moon Mimas, which is only 250 mi in diameter, had an enormous crater 80 mi wide with a huge mountain in the center, looking like an alien's giant eye. Scientists are still studying the pictures taken by the Voyagers, and by improving their quality using a computer technique called enhancement, are finding new things about the planet.

Early next century, a new spacecraft called Cassini may orbit the planet and send a small capsule to land on the surface of Titan; no spacecraft has yet landed on the moon of a planet other than the Earth's Moon.

SATURN'S RINGS

The rings circling Saturn consist of hundreds of "ringlets", which are kept in place by the gravitational force of small moons called "shepherd moons". This photograph of Saturn's rings taken from a Voyager spaceprobe has been computer enhanced to reveal possible variations in the chemical composition between different parts of the ring system.

URANUS

It is possible to see all the planets of the Solar System from Mercury to Saturn with the naked eye, so these planets have been known since ancient times. Uranus was the first planet to be found with a telescope, by the English astronomer Sir William Herschel in 1781. By comparing the position of this faint point of light against a star map, and noticing how it moved in relation to the stars, he concluded that he had discovered the seventh planet. It had been seen by earlier astronomers, but they had not noticed its slight movement.

Uranus is an unusual planet because its polar axis points not upward and downward, like the other planets', but sideways, so its North Pole is 98° from the vertical. The cause of this is unknown, but it has been suggested that the impact of a very large body could have been responsible. As a result of this extreme tilt, Uranus has 42 years of daylight, followed by 42 years of darkness. It also rotates in a counterclockwise direction. In 1977, astronomers discovered that Uranus had a ring system. There appeared to be nine tiny rings about 28,000 mi from the center of the planet. The central core of Uranus consists of molten rock and metal. This is surrounded by a deep atmosphere of mainly hydrogen and helium. Traces of methane are also present so, through a telescope, Uranus looks green as this gas absorbs red light. Uranus is 32,187 mi in diameter at its equator but, like the other large gaseous planets, bulges slightly at the equator

SEVENTH PLANET FROM THE SUN • AVERAGE DISTANCE FROM SUN 1,782 MILLION MI

Left: This is a montage of photographs taken by US spacecraft Voyager 2. It shows the planet (center) and the five of its 15 moons that can be seen using a telescope on Earth. These, clockwise from bottom left, are: Ariel, Umbriel, Oberon, Titania, and Miranda.

Top right: Launched in 1979, Voyager 2 reached Uranus in 1986.

Bottom right: Voyager 2 photograph of Miranda, showing the surface terrain. It has grooved and mottled areas, many craters and a strange V-shaped feature.

because of the speed of its rotation so the polar diameter is slightly less. Uranus takes 84 years to orbit the Sun at an average distance of 1,782 million mi. It appears as a featureless green disk through even the largest telescopes, however telescopic observations reveal five moons: Miranda, Ariel, Umbriel, Titania, and Oberon.

Voyager 2 Visits Uranus

Only one spacecraft has visited the planet, changing our image of Uranus in the process. This was Voyager 2, which was launched in 1979, and had already flown past Jupiter in July 1979 and Saturn in August 1981 before reaching Uranus in January 1986. Voyager discovered another 10 moons, all inside the orbits of the five known moons. Some of these new moons act as "shepherds" for the rings to keep them in place, of which there seemed to be 11 not nine. Unlike Saturn, whose rings are formed mainly of ice particles, those of Uranus appear to be made up of boulders.

The surface of the upper atmosphere of Uranus was disappointing in the Voyager photographs, as no clear bands of clouds could be seen but the spacecraft's images of the moon Miranda were spectacular. This is the closest and smallest of Uranus' major moons. Miranda, which is about 340 mi in diameter, seems to have every geological feature imaginable. There are strange areas of parallel grooves and ridges and sheer cliffs about 10 mi high. Scientists think that Miranda was torn apart a number of times and fused together in a confused geological muddle!

CIRCLES SUN IN 84 YEARS • DIAMETER 32,187 MI • TURNS ON ITS AXIS IN 15.5 HOURS

NEPTUNE

VOYAGER'S DISCOVERIES

The US spaceprobe, Voyager 2, measured wind speeds of 1,500 mi/h around Neptune (above) – the fastest in the Solar System. Neptune's two major rings were photographed from a distance of 620,000 mi by Voyager 2 (tinted pale blue, left). These narrow rings have radii of 33,000 mi and 39,000 mi and are composed of microscopic dust particles.

EIGHTH PLANET FROM THE SUN • AVERAGE DISTANCE FROM SUN 2,794 MILLION MI

This bluish ball of gas, about 30,750 mi in diameter, is the eighth planet in the Solar System but presently is also the furthest from the Sun! The reason for this is that the orbit of the furthest planet, Pluto, comes within the orbit of Neptune during a period of 20 years. The latest of these 20-year periods began in 1979 and will end in 1999.

The existence of Neptune had been predicted by many astronomers before it was actually seen. After Uranus was discovered, astronomers found, it was not keeping to its orbit around the Sun. They suspected that it was being pulled off course by the gravity of another, as yet unknown, planet. Mathematicians calculated where this new planet could be found, and, in 1846, the planet was discovered close to the predicted spot by the German astronomer Johann Galle.

Like the other giant planets, Neptune consists mainly of hydrogen and helium. It was thought to have had two moons before it was visited by Voyager 2. These moons are called Triton and Nereid. Nereid has an unusual, highly elliptical orbit, which takes it to within 800,000 mi of the planet, and as far out as 6,000,000 mi. It takes nearly a year to make one orbit. Triton is over 1,900 mi diameter and is one of the largest moons in the Solar System. It has a circular orbit, but orbits from east to west. Scientists believe that an extraordinary event may have caused the unusual orbits of these two moons. They think that Pluto may have been a moon of Neptune that wandered too close to Triton and was thrown out of orbit like a pool ball. The gravitational effect of this encounter disturbed the orbits of Nereid and Triton.

Voyager 2 Visits Neptune

In August 1989, after flying past Jupiter, Saturn and Uranus, Voyager 2 flew past the fourth planet of its 10-year voyage. Neptune's swirling atmosphere, with "scooter" clouds that rotate faster than the rest of the clouds, was much more interesting than that of Uranus. Four new moons were discovered, plus the suspected ring system. Photographs of the moon Triton were spectacular. Through the thin atmosphere, an icy, swampy surface of nitrogen and methane could be seen, as well as volcanoes that spew liquid nitrogen a distance of 25 mi at a speed of 820 ft/sec, like gigantic geysers.

Above: Voyager 2 leaves Neptune's moon, Triton, in August 1989 to travel out of the Solar System. In 368,000 years, it will pass to within 0.8 light years of Sirius, the brightest star seen from Earth and fifth closest star to the Sun.

CIRCLES SUN IN 164.8 YEARS • DIAMETER 30,750 MI • TURNS ON ITS AXIS IN 18 HOURS

PLUTO

Left: An artist's impression of Pluto. It was discovered in 1930 by comparing photographs of the same area of the sky. A moon, Charon, was not discovered until 1978.

Below: The surface of Pluto must be very cold indeed, probably about −380 °F. This is because the Sun is 1,000 times fainter than on Earth.

NINTH PLANET FROM THE SUN • AVERAGE DISTANCE FROM SUN 3,666 MILLION MI

Pluto is a tiny planet; it is the smallest in the Solar System. A globe of frozen gases, most probably water, ammonia, and methane, it measures about 1,500 mi in diameter. It takes Pluto 248 years to make one orbit around the Sun. Its average distance from the Sun is 3,666 million mi but it has an elliptical orbit which takes it as close as 2,700 million mi and as far as 4,500 million mi. At the moment, it is closer to the Sun than Neptune is, because since 1979 it has been orbiting within Neptune's orbit, and will continue to do so until 1999.

Because it is so small, Pluto was only discovered by studying photographs of an area of the sky and noticing that a small speck of reflected light moved very slightly against a background of static stars. Pluto's position had been predicted for several years, but it was the American astronomer Clyde Tombaugh who discovered it on photographic plates in 1930. Pluto is the last planet to have been discovered. Because it is so tiny and so far away, very little is known about it; even the most powerful telescopes show it only as a star-like point of light.

The Moon of Pluto

Before 1978, the only known facts about Pluto were that it apparently rotated once every 6.3 days and had a probable surface temperature of $-382°$F. In 1978 came the dramatic discovery that Pluto had a moon. It was called Charon. Charon is about one-third the size of Pluto, with a diameter of about 500 mi. It makes one orbit around Pluto every 6.3 days; this is the same time that Pluto takes to rotate on its axis, so Charon is always above the same place on the planet. Astronomers believe that Pluto and Charon were once moons of Neptune that broke out of its orbit.

CIRCLES SUN IN 248 YEARS • DIAMETER 1,500 MI • TURNS ON ITS AXIS IN 6.3 DAYS

COMETS

When the planets in the Solar System were formed, pieces of material, mainly rocks and dust, were left orbiting the Sun. In the icy reaches of deep space, much of this material froze together, with ice and gas, to form "dirty snowballs", about 3,300 ft across. These frozen bodies loop the Sun in a highly elliptical orbit. This means that they travel very deep into the Solar System and come relatively close to the Sun.

As they approach the Sun, the dirty snowballs start to warm up and expand. The heat causes the gases to evaporate and dust is released, forming a fuzzy head and a long, flowing tail. When this happens, these bodies are the brightest and largest objects in the sky. They are called comets. They blaze brilliantly for several weeks or months and then head back into space.

Halley's Comet

The most famous comet is named after astronomer Edmond Halley, who, in the early eighteenth century, noted that the comets seen in 1682, 1607, and 1531 were remarkably similar. He believed that they were all in fact the same comet and was proved correct. Halley's Comet (as it is now known) orbits the Sun every 76 years on average. The comet's path takes it from about 55 million mi from the Sun (between the orbits

Left: Comet Webb as it passed close to the Sun in 1976, showing a broad, fin-shaped dust trail above and a narrow, blue gas trail below.

INSIDE A COMET

A comet consists of two main parts: the head and the tail. In the center of the comet's head is a small potato-shaped, rocky body – the nucleus. It is made of dust and ice, and is the only solid part of a comet. As the comet approaches the Sun, the nucleus releases the gas and dust to form a glowing sphere – this is the coma. Radiation from the Sun, called solar wind, carries gas and dust from the coma into a tail that streams away from the comet's head. Comet tails always point away from the Sun. The comet's gases glow brightly, making the comet visible to the naked eye.

ENCKE'S COMET HAS SHORTEST KNOWN ORBIT • ORBITS THE SUN EVERY 3.3 YEARS

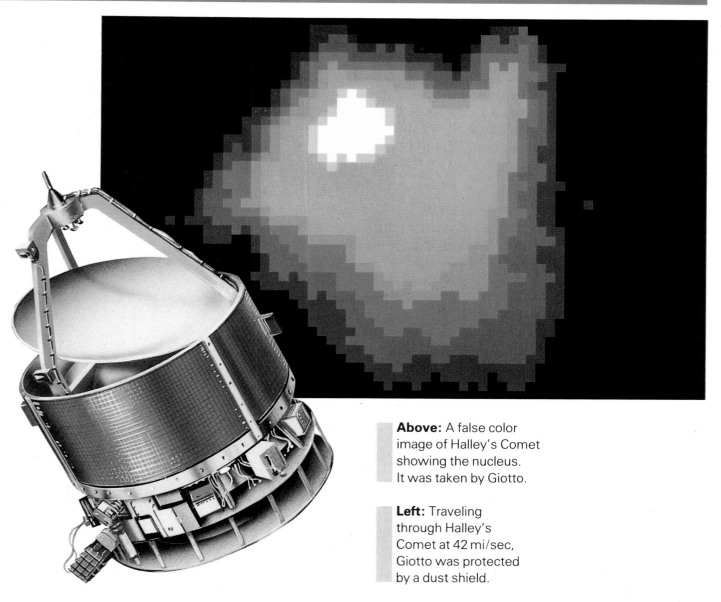

Above: A false color
image of Halley's Comet
showing the nucleus.
It was taken by Giotto.

Left: Traveling
through Halley's
Comet at 42 mi/sec,
Giotto was protected
by a dust shield.

of Mercury and Venus) to over 3,000 million mi from
the Sun (beyond the orbit of Neptune). It was first
recorded in 86 BC. The comet appeared in 1066 shortly
before the Battle of Hastings, and is shown on the
Bayeux Tapestry. After another appearance in 1301,
the Italian artist Giotto di Bondone depicted it in his
painting *The Adoration of the Magi*. In 1986, Halley's
Comet was the first comet to be explored at very close
quarters by a spacecraft. The European spaceprobe
Giotto actually flew through the comet's head at a very
high speed and was battered by small particles of dust.

Halley's Comet is one of about 400 comets which
return to the Sun within every 200 years. There are
about 500 other comets that are in such deeply elliptical orbits that they will not return for thousands of
years. Unfortunately, most of the comets that appear
each year are too faint to be seen with the naked eye.
However, 10 or more may be visible through a telescope, and anyone who discovers a new one has it
named after them!

HALLEY'S COMET ORBITS THE SUN EVERY 76 YEARS • NAMED AFTER EDMOND HALLEY

METEORS

Meteors are often called shooting stars, which is exactly what they look like. They are streaks of light caused by a solid particle from space entering the Earth's atmosphere at a height of about 60 mi and at a speed of about 30 mi/sec. They burn up as a result of atmospheric friction into a trail of bright dust. It is possible to see about 10 meteors an hour on a clear night. They appear as sudden streaks of light that dart across part of the sky, usually vanishing in a second or less.

Some meteors come in showers. These are caused by dust shed from a comet. Their appearance in the sky is very predictable and is often associated with the constellation in which the display occurs. For example, debris from Halley's Comet hits the Earth's atmosphere every year during May and October producing spectacular meteor showers. These displays of perhaps 20 meteors per hour are called the Eta Aquarids and Orionids, as they appear in the constellations of Aquarius and Orion. The best meteor shower is the Quadrantids, which can average over 100 sightings per hour. These appear during the first days of a new year.

Meteorites

Sometimes a large lump of rock or metal can survive the entry into the Earth's atmosphere and hit the ground. This is known as a meteorite. Meteorites are fragments of asteroids – debris left over from the formation of the planets. About 500 meteorites can hit the Earth every year. If a meteorite is moving quickly enough when it hits the ground, it can blast out a gigantic crater. Fortunately, most land in the oceans or in the vast uninhabited regions of the Earth. Meteorites are usually very small, rather like stones, but some are extremely large.

The largest known meteorite was found at Grootfontein, in Namibia, south-west Africa in 1920. It is about 9 ft long by 8 ft wide, and weighs 60 t. A meteorite weighing about 2 million t and measuring about 260 ft in diameter hit the Earth in about 25000 BC, causing the huge Coon Butte crater in Canyon Diablo, near Winslow, Arizona. This must have caused an explosion equivalent to a huge nuclear bomb. It left a crater 4,150 ft in diameter and 575 ft deep. However, nothing remains of the meteorite.

Below: Is this what happened in 25000 BC? A 2-million t meteorite, about 260 ft wide, crashes to Earth in what is now Coon Butte, in Canyon Diablo, Winslow, Arizona, USA.

LARGEST KNOWN METEORITE 9 X 8 FT, WEIGHING 60 T, FOUND IN NAMIBIA IN 1920

Right: Aerial photographs of the Winslow meteor-crater (top), and a similar one at Gosses Bluff in the Northern Territory of Australia (bottom).

LARGEST METEORITE CRATER IS THE COON BUTTE, 4,150 FT DIAMETER, 575 FT DEEP

ROCKET POWER

If you blow up a balloon, then let it go, it flies wildly around the room. This is caused by reaction to the thrust of the air rushing out of the balloon's neck. A rocket works in just the same way and it was by developing rocket power that artificial Satellites became a practical proposition.

In a rocket engine, propellents, called fuel, and an oxidizer (which supplies the oxygen to burn the fuel), are ignited (set on fire) to form hot gases (exhaust), which are forced through a nozzle at the end of the engine. It is this thrust that propels the rocket. Unlike the balloon, the rocket's flight path is controlled by gyroscopes and computer guidance systems to make sure it keeps heading in the right direction.

Most modern rockets use liquid fuel, such as kerosene or liquid hydrogen, which is burned with liquid oxygen to produce the combustion and thrust. Liquid hydrogen fuel and liquid oxygen oxidizer are called cryogenic propellents because they are gases that have been cooled so much that they have turned into a liquid. The propellents are pumped from storage tanks into a chamber, where they are mixed and ignited. Some propellents, such as hydrazine and nitric oxide, are called hypergolic because they ignite as soon as they come into contact, in an explosive chemical reaction.

The Space Shuttle is unusual in that it uses both liquid and solid propellents. The three main engines on the Shuttle itself use liquid oxygen and liquid hydrogen, while the two big white booster rockets either side of it are propelled by a solid propellent, which is a mix of chemicals in a kind of rubbery substance.

The Multi-Stage Rocket

To carry an artificial satellite into orbit, a rocket must reach a speed of about 17,900 mi/h so it is fast enough to overcome the pull of the Earth's gravity. One rocket cannot do this on its own, so several fuel tanks are joined together; each section is called a stage. The stages are stacked one on top of the other or added at the side of the rocket. Most space rockets have three stages, each one igniting in turn to increase the speed gradually; they separate and fall away as the fuel runs out; this makes the rocket much lighter.

The Space Shuttle has a large tank which holds the fuel for its engines, and two solid-fuel rocket boosters

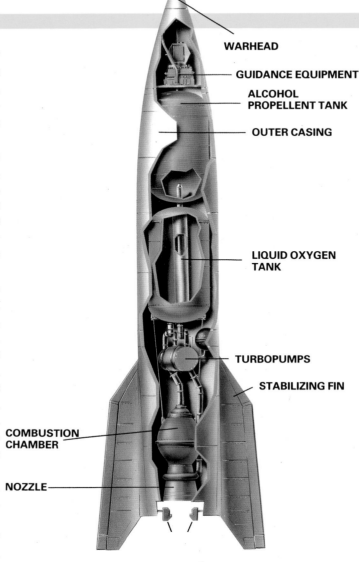

WARHEAD

GUIDANCE EQUIPMENT

ALCOHOL PROPELLENT TANK

OUTER CASING

LIQUID OXYGEN TANK

TURBOPUMPS

STABILIZING FIN

COMBUSTION CHAMBER

NOZZLE

Above: One of the first large rockets was the German V2, which was used as a weapon during World War II.

are strapped to the sides of the fuel tank. These give an extra boost during the first two minutes of flight, and then parachute back to Earth to be used again. The liquid propellent engines burn for another six minutes before the Shuttle reaches orbit, then the empty fuel tank falls away and breaks up. Smaller rockets on the Shuttle itself are sometimes needed for the final boost to the planned orbit.

GODDARD BUILT AND FIRED FIRST LIQUID-FUELED ROCKET ON MARCH 16 1926

Right: The world's tallest rocket, the American Saturn V. Standing 364 ft tall, it weighed 3,000 t on the launch pad. The three-stage Saturn V launched the Apollo crews to the Moon between 1969 and 1972.

THIRD STAGE FIRES TO THRUST THE APOLLO SPACECRAFT INTO ORBIT THEN SEPARATES

SECOND STAGE FIRES, BOOSTS THE APOLLO SPACECRAFT HIGHER AND FASTER, THEN SEPARATES

STAGE SEPARATION SKIRT

Left: The staging sequence of Saturn V with a first stage thrust of 7.7 million lb. Earth orbit is reached 11 mins after blast off from the Kennedy Space Center, Florida.

FIRST STAGE BURNS OUT AND FALLS AWAY

SATURN V, WORLD'S MOST POWERFUL ROCKET, 364 FT TALL, CARRIED 3,000 T OF FUEL

SATELLITES

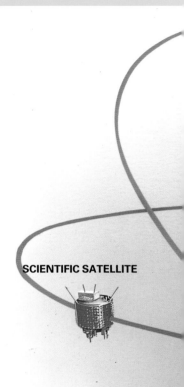

SCIENTIFIC SATELLITE

SATELLITE DESIGN

Because satellites only travel in space, they need not be streamlined, but can be whatever shape is best for the job they have to do. The body of a satellite is usually made of ultra-lightweight aluminium alloys and strong plastics. On it are all kinds of equipment, such as cameras, telescopes, detectors, and many different measuring instruments which stick out into space. Telstar (left) was the first communications satellite, launched in 1962.

Artificial satellites are the most common kind of spacecraft, and they stay in space for many years. They have a variety of uses: communications, astronomy, and observation of the Earth and its atmosphere from space.

A satellite reaches orbit when it is traveling at a speed that is fast enough to overcome the force of the Earth's gravity, which is trying to pull it back to Earth, while at the same time, it is not fast enough to make the satellite speed away from Earth's gravity altogether.

About 5,000 artificial satellites have been launched into Earth orbit since Sputnik 1 in 1957. They were not however all sent into the same orbit. In a typical orbit, the craft will travel around the Earth at about 250 mi altitude. At this height, it will be traveling at about 17,900 mi/h in a circular orbit. If the satellite is in a much higher orbit, its speed need not be as great because the Earth's gravity becomes weaker.

Most satellites are sent into elliptical orbits which have a high point (apogee) and a low point (perigee) above the Earth.

Different Orbits

Satellites go into orbits which travel around the Earth in different directions. For example, a Space Shuttle may orbit so that it crosses the equator at an angle of 28°. This is called the orbital inclination. It means that during its orbits, the Shuttle never reaches further north or south of the 28° latitude on the Earth. A satellite that orbits at 90°, however, flies over the poles of the Earth and therefore covers all latitudes. If it travels in a low orbit around the poles, it takes 90 minutes to go around once, and in one day it makes 17 orbits while the Earth has rotated once, so the satellite flies over every area of the Earth in this period. This is an important orbit for some observation satellites, particularly those that spy!

Another very important orbit is called the geostationary orbit. Here the satellite enters a circular orbit at an altitude of about 22,300 mi where it takes 24 hours to make one orbit, which is the same time it takes the Earth to rotate on its axis. The satellite also orbits at 0° inclination, directly above the equator, so it seems to be stationary, hovering over the same spot on the Earth. This is the perfect situation for a communications satellite which is rather like a TV and radio antenna in the sky.

IN A CIRCULAR ORBIT OF 250 MI, SATELLITES TAKE 90 MINUTES TO CIRCLE EARTH

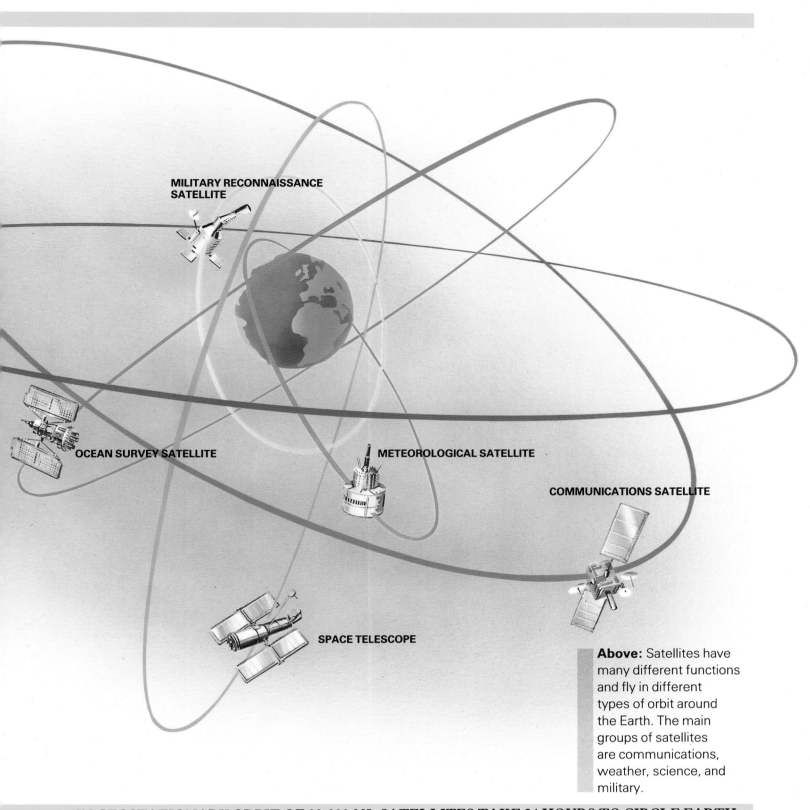

MILITARY RECONNAISSANCE SATELLITE

OCEAN SURVEY SATELLITE

METEOROLOGICAL SATELLITE

COMMUNICATIONS SATELLITE

SPACE TELESCOPE

Above: Satellites have many different functions and fly in different types of orbit around the Earth. The main groups of satellites are communications, weather, science, and military.

IN GEOSTATIONARY ORBIT OF 22,300 MI, SATELLITES TAKE 24 HOURS TO CIRCLE EARTH

POWER FOR SPACECRAFT

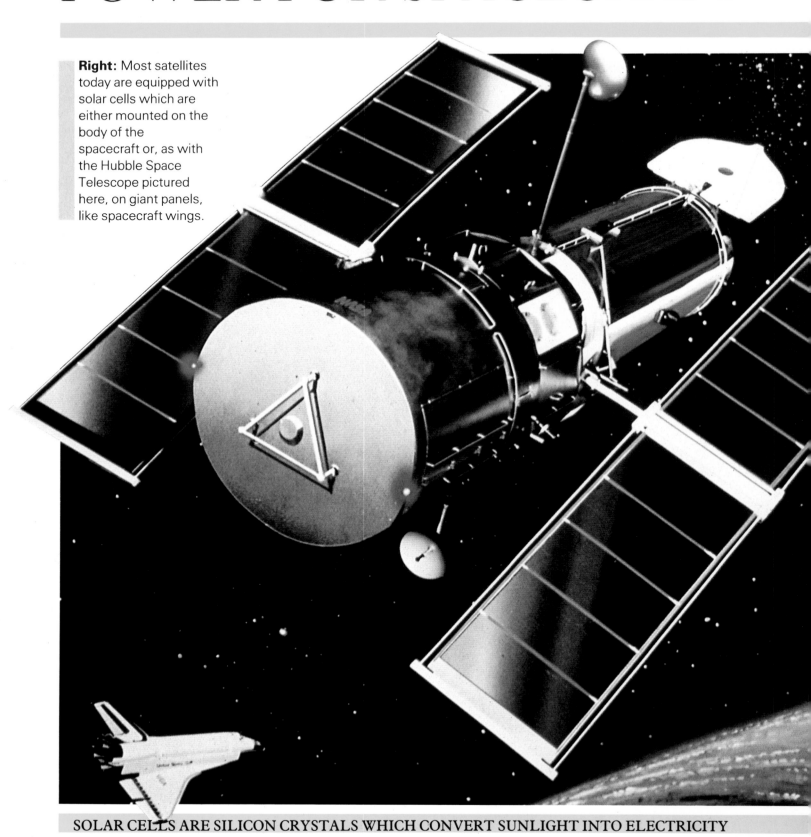

Right: Most satellites today are equipped with solar cells which are either mounted on the body of the spacecraft or, as with the Hubble Space Telescope pictured here, on giant panels, like spacecraft wings.

SOLAR CELLS ARE SILICON CRYSTALS WHICH CONVERT SUNLIGHT INTO ELECTRICITY

The first artificial Earth satellite, Sputnik 1, was powered by a chemical battery. However, after a few days the battery ran out of power and the satellite's radio transmitter went dead. In 1958, America launched the satellite Vanguard 1, which tested a new type of system powered by solar cells. These are tiny pieces of silicon glass that absorb the Sun's light and convert this energy into electrical power.

Solar Energy

Thousands of solar cells mounted on spacecraft can provide 1,500 watts of electricity or even more. Most satellites today are powered this way. Some satellites have giant wings, and the solar cells are mounted on these solar panels. Other satellites have their entire bodies covered with solar cells. Scientists are studying the use of more productive solar cell materials, such as gallium arsenide crystals.

Fuel Cells

There are other types of electrical generators on spacecraft. The Space Shuttle uses fuel cells. These are provided with oxygen and hydrogen propellents from liquid gas tanks on the Shuttle. The cells then convert the chemical energy of the reaction between the two gases into electrical energy. A by-product of this fuel cell process is drinkable water.

Radioactive Power

Some satellites which need very large amounts of electricity, such as spacecraft using large radar scanners to observe the Earth in great detail, use radioactive power sources. These spacecraft have a radioactive fuel core of uranium or plutonium. As the radioactive material decays, it produces energy which is then converted to electricity. Spacecraft that travel into the distant Solar System where there is no sunlight to power solar cells use this radioactive fuel system, called a radioisotope thermoelectric generator or RTG. The Voyager 2 spacecraft, which was launched in 1977 and traveled to Jupiter, Saturn, Uranus, and Neptune, reaching Neptune in 1989, could not have operated had it not been for its plutonium-powered RTG.

The USA has plans to power space stations of the future with huge solar mirrors. These concentrate the Sun's heat and use this heat, or thermal energy, to drive generators which provide electricity.

SPACEPROBES TRAVELING INTO DEEP SPACE ARE POWERED BY NUCLEAR BATTERIES

COMSATS

TV SIGNALS FROM AN OUTSIDE BROADCAST TRUCK
TRAVEL BY CABLE TO A MOBILE RECEIVER,
WHICH TRANSMITS THE SIGNALS
TO A COMMUNICATIONS SATELLITE

Above: Communications satellites are used to relay TV direct to homes; transmit telephone, telex, fax, and computer data; and to keep mobile vehicles in touch with headquarters.

TELSTAR WAS FIRST COMMUNICATIONS SATELLITE, LAUNCHED BY USA IN 1962

In 1962, the USA launched the world's first communications satellite, called Telstar. It could receive television pictures and relay them live across the Atlantic Ocean. This meant that when an American astronaut called Walter Schirra was launched into the Earth's orbit aboard Mercury 8 later that year, TV viewers in Britain were able to see his launch the same day via the Telstar communications satellite. Before this, a film of the event would have been sent by plane, so that it could not have been seen until the next day. Today, most long-distance communications: TV, telephone, telex, fax (facsimile transmission), and computer data go through communications satellites, or comsats.

An international organization called INTELSAT (International Telecommunications Satellite Organization) was set up in 1964 to provide worldwide communications for commercial use. Today there are over 100 member countries, and they own and operate the world's largest and most powerful commercial communications satellite called Intelsat 6. The first satellite in the Intelsat 6 series was launched in 1989. Intelsat 6 is shaped like a garbage pail, is 40 ft high and has several dishes attached to its top. These are the communications antennas. In its stationary orbit, 22,300 mi above the equator, Intelsat 6 relays 30,000 simultaneous telephone calls between two continents, operates three television relay channels at the same time, and transmits 3 billion bits of computer information per second.

How a Communications Satellite Works

A communications satellite collects electronic information – such as voice, data, and TV pictures – from a transmitting station on the ground, increases the power of these signals by amplifying them, then retransmits the signals to ground stations on Earth. The TV satellites which provide programs directly to homes equipped with small satellite dishes operate on very high power, enabling the signal to be collected by small receivers.

Communications satellites need a lot of electricity to operate. Intelsat 6, for example, requires 2,400 watts of power, and this power is provided by solar cells mounted around its enormous cylindrical body.

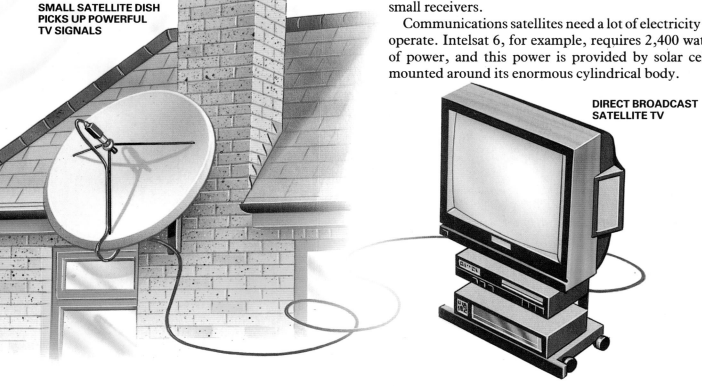

SMALL SATELLITE DISH PICKS UP POWERFUL TV SIGNALS

DIRECT BROADCAST SATELLITE TV

INTELSAT 6, 40 FT LONG, CAN HANDLE 30,000 TWO-WAY CALLS AT THE SAME TIME

WATCHING THE EARTH

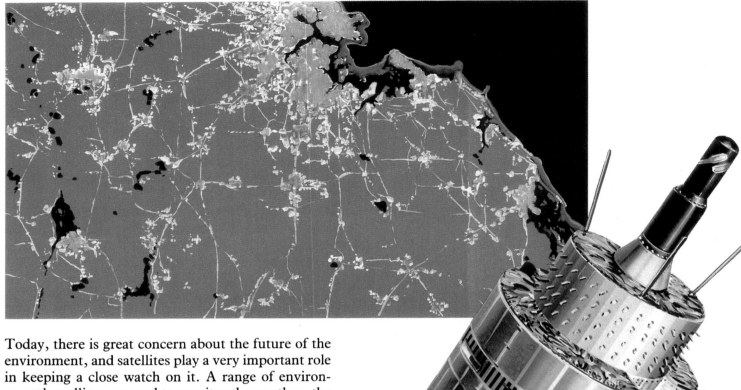

Today, there is great concern about the future of the environment, and satellites play a very important role in keeping a close watch on it. A range of environmental satellites are used to monitor the weather, the temperature of the land and sea, wind speed, the height of clouds, the speed of ocean currents, the location of minerals and water on land, the movement of animals, industrial pollution, deforestation of the jungles, and even the hole in the ozone layer of the Earth's atmosphere.

Environmental Satellites in Orbit

There are two types of environmental satellite. A satellite in a stationary orbit 22,300 mi above the Earth returns images of one-third of the face of the Earth every few minutes. Three of these satellites strategically located at points over the equator can therefore observe the whole Earth. A polar orbiting satellite covers the whole globe in a day from a low orbit (about 800 mi above the Earth), orbiting 17 times in one day, as the Earth rotates beneath it. Many of these satellites also carry receivers which can pick up distress signals from people lost at sea or on land.

The Work of Environmental Satellites

Environmental satellites are fitted with many types of instruments to monitor aspects of the environment

Top: A false color image from the US spacecraft, Landsat 5, showing the city of Boston. The red color is lush vegetation.

Above: Meteosat, a European weather satellite which is in geostationary orbit around the Earth.

ENVIRONMENTAL SATELLITES MONITOR THE EARTH'S ATMOSPHERE AND SURFACE

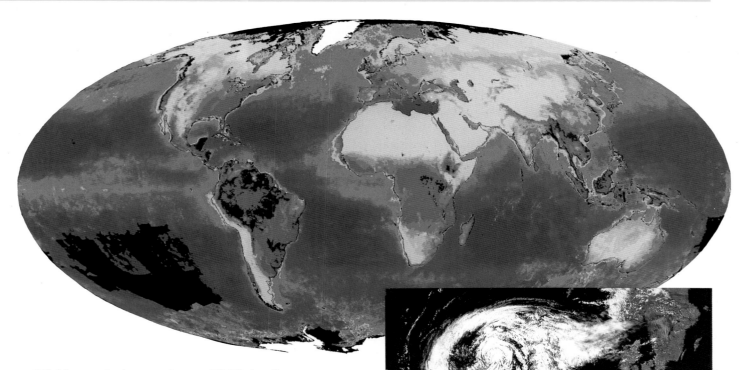

Above: An image of the Earth's vegetation showing details such as the level of plankton (microscopic plants) in the sea (red and yellow), and the rain forests (light green).

Right: Europe as seen from the US weather satellite NOAA 9, which orbits around the Earth's poles. A severe storm system covers most of the British Isles.

and can often spot the hidden details. Some send back photographs which show the Earth as we would see it, but photographs are also taken with infra-red cameras. These can "see" through cloud and detect infrared (heat) radiation, so the images show how hot or cold parts of the sea or land are.

Other instruments identify chemicals in the atmosphere; these are used to monitor pollution, the greenhouse effect, and the hole in the ozone layer. Images of oil slicks and industrial waste polluting the water can also be produced.

Some imaging systems use radar waves to make up pictures of the Earth, particularly to map the surface of the Earth under the sea. Photographs taken by satellites have even been used to help geologists find rare minerals in remote parts of the world. Radar pictures have also taken a peep into history: Space Shuttle images showed the remains of a civilization under the sand in Africa.

TIROS 1, LAUNCHED BY THE USA IN APRIL 1960, WAS THE FIRST WEATHER SATELLITE

61

MILITARY SATELLITES

Over half the satellites sent into Earth orbit by the USSR and the USA are used for military purposes: to observe enemy territory. Some take close-up photographs of targets on the ground, while other satellites are used for eavesdropping on radio communications and defense radars. Some military satellites have more specialized purposes, such as monitoring enemy launch sites to give early warnings of missile attack, or use radar and other sensors to track enemy ships and submarines. There are also military communications satellites which are used to link the leader of a country with a troop commander in the battlefield.

The first military satellites were called Discoverer and were launched by the USA 30 years ago. These small capsules held cameras which took photographs of locations in the USSR. The capsules returned to Earth and were caught in mid-air by aircraft trailing a snare-net, rather like a fishing net. The film was processed and the photographs analyzed.

Today, such recoverable spacecraft are still used. They use camera systems that are powerful enough to locate an object on Earth about 10 ft wide from an orbit about 190 mi high. The main disadvantage of these spacecraft is that the film has to be recovered and processed, and this can take valuable time.

Modern Spy Satellites

Spy satellites have been developed which electronically scan the ground from orbit and transmit the images to Earth by radio. These can now return images with as good a clarity, or resolution, as film cameras. They are also able to maintain a continuous watch without running out of film.

Electronic monitoring, or elint, satellites carry huge dish antennas which pick up all sorts of radio transmissions. This radio jumble of noise can then be transmitted to a ground station and filtered, so that important and perhaps secret conversations can be bugged.

Other military satellites are used to monitor the weather for naval, air, and land operations, and others are used to maintain secret communications. Communications are so good today that a soldier with a small radio back-pack can talk via satellite with headquarters on the other side of the world. Similar, small satellite receivers enable soldiers, naval commanders and pilots, to navigate with extreme precision, using several navigation satellites.

STAR WARS

In 1983, President Reagan announced his Strategic Defense Initiative or Star Wars plan to put high energy lasers into orbit to shoot down incoming enemy missiles. The results of this research (left): a Titan missile stage is destroyed by a chemical laser in a test at White Sands, New Mexico. The test was to see how vulnerable a liquid-fueled rocket is to an attack. The rocket stage was stressed to simulate exact flight conditions.

WITH MILITARY SATELLITES, USA AND USSR KNOW THE ARMAMENTS EACH SIDE HAS

Above: Cape Canaveral is the triangular sand spit at the bottom of this photograph taken from the Space Shuttle. The Kennedy Space Center is above right, where the Shuttle runway and two launch pads can be seen.

Left: A Delta booster is launched from Cape Canaveral carrying a Star Wars experiment into orbit. Several Star Wars tests are planned before the system is in place.

FIRST US MILITARY SATELLITES, CALLED DISCOVERER, WERE LAUNCHED IN 1959

THE SPACE RACE

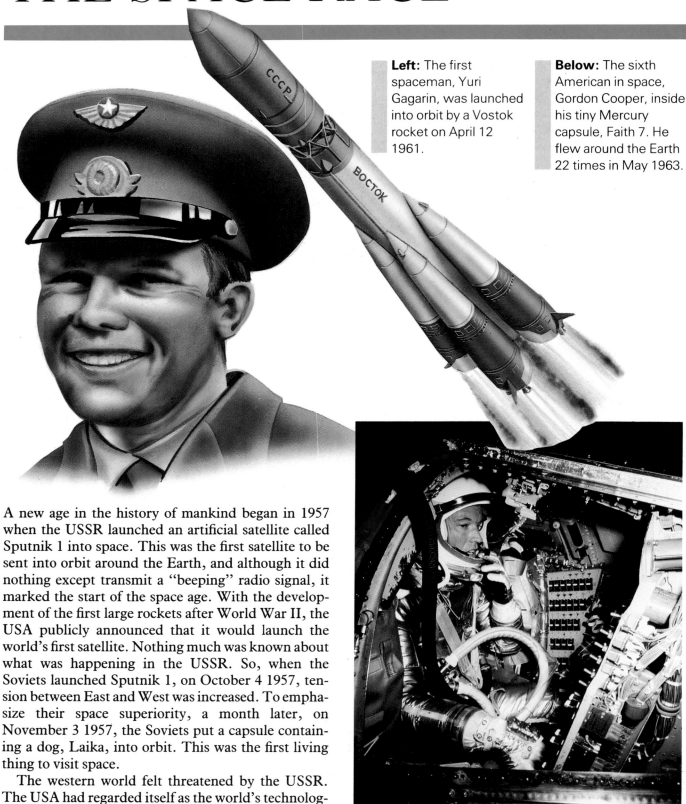

Left: The first spaceman, Yuri Gagarin, was launched into orbit by a Vostok rocket on April 12 1961.

Below: The sixth American in space, Gordon Cooper, inside his tiny Mercury capsule, Faith 7. He flew around the Earth 22 times in May 1963.

A new age in the history of mankind began in 1957 when the USSR launched an artificial satellite called Sputnik 1 into space. This was the first satellite to be sent into orbit around the Earth, and although it did nothing except transmit a "beeping" radio signal, it marked the start of the space age. With the development of the first large rockets after World War II, the USA publicly announced that it would launch the world's first satellite. Nothing much was known about what was happening in the USSR. So, when the Soviets launched Sputnik 1, on October 4 1957, tension between East and West was increased. To emphasize their space superiority, a month later, on November 3 1957, the Soviets put a capsule containing a dog, Laika, into orbit. This was the first living thing to visit space.

The western world felt threatened by the USSR. The USA had regarded itself as the world's technolog-

SOVIETS LAUNCH SPUTNIK 1 1957 • FIRST SPACE FLIGHT BY YURI GAGARIN 1961

ical leader and now found itself outpaced by the Soviet leap into space. Top priorities were given to space activities. America's first attempt to place a satellite into orbit failed when the rocket blew up on the launch pad. It was not until January 1958 that Jupiter C launched the first US satellite Explorer I into orbit. The Soviets continued to send up Sputniks and new Luniks. Luna 2 was the first spacecraft to reach the Moon in 1959, and Luna 3 was the first satellite to photograph the far side of the Moon.

America suffered many failures, which were seen all over the world because of the open nature of its program. The USSR suffered failures too, but the news never leaked out, so it formed an invincible lead in what had become the space race.

In August 1960, the Soviets sent dogs into orbit for one day to test the new Vostok spacecraft in which Soviet cosmonauts would eventually fly; it was disguised under the name Sputnik 5. The two dogs, Belka and Strelka, were the first living things to return from orbit.

The First Men in Space

The next step was to launch the first man in space. On April 12 1961, Yuri Alekseyevich Gagarin was rocketed into space and, after completing one orbit, the automatic controls of his spacecraft brought him safely back to Earth. The first American was sent into space on May 5 1961, when Alan Shepherd made a simple up-and-down sub-orbital flight in a Mercury capsule. The first American to orbit Earth was John Glenn in 1962. The Soviets launched a new spacecraft and, in March 1965, Alexei Leonov crawled through the airlock of Voskhod 2 to make the first walk in space.

The Moon Race

In 1961, President Kennedy had set the USA the target of landing a man on the Moon by 1970 – before the USSR. Project Gemini in the mid-1960s pioneered most of the systems that would be needed for the trip to the Moon in the Apollo spacecraft. Space-walking (to perform tasks outside the capsule in space), rendezvous and docking of spacecraft, maneuvering in space and spending enough time in space (14 days) for the lunar trip were all tried out on the 12 Gemini missions.

Above: Ed White was the first American to walk in space on June 3 1965. A cable connected him to Gemini 4 and supplied him with oxygen.

The Apollo project was the final stage of the US program to place a man on the Moon and bring him back safely to Earth. After a number of test flights, first around the Earth and then around the Moon without landing, a giant Saturn V rocket lifted Apollo 11 from Cape Canaveral in Florida on July 16 1969. On July 21, Neil Armstrong stepped onto the surface of the Moon.

ALEXEI LEONOV MAKES FIRST WALK IN SPACE 1965 • FIRST MAN ON MOON 1969

MAN ON THE MOON

On July 21 1969, Neil Armstrong climbed down the ladder of the lunar lander and onto the surface of the Moon. Placing his foot on the lunar soil, he said, "That's one small step for man, one giant leap for mankind." He was the first person to set foot on the Moon. Buzz Aldrin then joined Armstrong on the Moon's surface. They spent over two hours doing experiments, and collected over 44 lb of rock and soil. Then they returned to the spacecraft and blasted off to rejoin the third astronaut, Michael Collins, who had remained orbiting the Moon while they visited it.

The Apollo Spacecraft

Armstrong, Aldrin, and Collins had traveled to the Moon aboard Apollo 11. This was shot into space by the most powerful rocket ever built, Saturn V. The Apollo spacecraft consisted of three parts. The command module was a three-man pressurized capsule where the astronauts sat for take-off; it was also the only part of the spaceship to return to Earth. This was attached to the service module which contained fuel

Below: Apollo landings. 11: July 1969, Sea of Tranquillity. 12: November 1969, Ocean of Storms. 14: February 1971, Fra Mauro. 15: July 1971, Hadley Rille. 16: April 1972, Descartes. 17: December 1972, Taurus Littrow.

Above: An Apollo lunar module on the Moon.

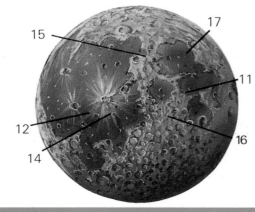

LIFT-OFF 9:32 AM WEDNESDAY JULY 16 1969 • MAN ON MOON 2:30 AM MONDAY JULY 21

and equipment, plus a rocket motor used to correct the course when traveling to and from the Moon. The third part of the Apollo spacecraft was the lunar module, which went down to the Moon's surface. The lunar module was abandoned before the flight home and the service module was thrown away just before the command module entered the Earth's atmosphere, making the Apollo program very expensive. It is estimated to have cost $25,000 million.

There were other manned expeditions to the Moon after Apollo 11. However, one mission, Apollo 13, went wrong. An oxygen tank exploded on the way to the Moon, so the landing was canceled, but luckily none of the astronauts was injured.

Exploring the Moon

Between December 1968 and December 1972, 23 people went to the Moon and 12 walked on the surface. During the Apollo 11, 12, 14, 15, 16, and 17 missions, 160 hours were spent exploring the surface at six different sites; they covered 60 mi. Apollo 15, 16, and 17 carried the lunar roving vehicle, or Moon buggy. This small, four-wheeled car was powered by electric motor and had a maximum speed of 10 mi/h. The astronauts collected 2,196 samples of the Moon, weighing a total of 850 lb, for analysis. They carried out experiments and set up scientific stations, one of which is still sending information back to Earth. They also took over 30,000 photographs.

Below: Apollo 15 commander Dave Scott salutes the US flag at Hadley Base.

12 APOLLO ASTRONAUTS LAND ON MOON BETWEEN JULY 1969 AND DECEMBER 1972

THE SPACE SHUTTLE

Nobody throws away an airplane after just one flight, so why throw away a rocket and a spacecraft? This was the logic behind the development of the Space Shuttle. In 1972, the USA decided to build such a shuttle. It would be in service in 1978 and be capable of making over 50 trips a year, making space travel almost as routine as airflight.

The first Space Shuttle went into space on April 12 1981. The main part of the Space Shuttle is the orbiter which looks like an ordinary airplane. It is launched like a rocket but flies back to land on a runway like an aircraft. The orbiter measures 122 ft long and has a wingspan of 78 ft. The forward compartment carries up to eight people, including scientists who carry out experiments in space. This is the only pressurized section of the orbiter. The upper flight deck looks very like that of an airplane only with more dials, controls, and instruments and at the rear of the upper deck are controls for launching satellites. The lower deck consists of a galley or kitchen, toilet, and sleeping quarters. Most of the orbiter's length is taken up by the huge payload bay which can carry a payload (cargo) of 29.5 t into orbit and bring 14.8 t back to Earth. Inside the payload bay is a remote-controlled arm used for handling satellites. At the rear of the orbiter are the three main engines used during launch; in space the Shuttle maneuveres using small rockets in the nose and tail.

The Space Shuttle is a remarkable flying machine. It is flown by five computers. Four are used to fly the Shuttle at any one time, with their decisions cross-checked to spot any errors. Using the Shuttle, it has been possible to fly up to a faulty satellite and repair it, or bring it back to Earth. It can carry three satellites at once, and place them into orbit from its payload bay. However, the Space Shuttle is not entirely reusable; it jettisons its empty 33-t fuel tank which burns up in the atmosphere just before it goes into orbit. Nor is it as safe or reliable as an airliner. In January 1986, the Space Shuttle exploded just after it was launched, killing seven astronauts and completely destroying the Shuttle. Only about ten launches a year are possible, not the 50 first hoped.

The Future of the Shuttle

In the future, the Space Shuttle will be used to carry parts of an international space station, called Freedom, into orbit, where Shuttle crews will assemble the space station helped by mechanical robot arms. The Shuttle will then be used to fly crews and equipment to operate the space station. The Shuttle will also be used to place huge telescopes into orbit which can be serviced by later Shuttle crews.

The US Shuttle is just the first of several that have been planned. Later versions may be launched by more powerful rockets and will therefore be able to carry more equipment. Eventually, an entirely new version, called Shuttle 2, may be built using new engines and materials technology to reduce the cost of going to and from space.

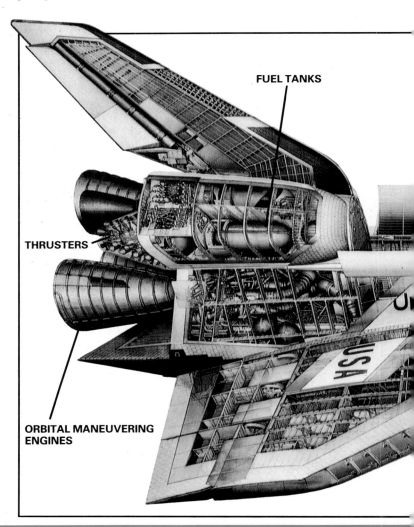

FUEL TANKS

THRUSTERS

ORBITAL MANEUVERING ENGINES

ORBITER 122 FT LONG, WINGSPAN OF 78 FT, WEIGHT 150,000 LB, FLOWN BY COMPUTER

Right: An artist's impression of the Space Shuttle delivering a payload into orbit from its payload bay. The payload comprises a satellite and a rocket stage attached to it, which will be used to carry it into a higher orbit. The satellite is lifted out of the payload bay by a remote-controlled arm operated by the astronauts. The arm is also used to recover satellites and return them to the hold.

PAYLOAD BAY

COCKPIT

SATELLITE CARGO

CREW QUARTERS

Left: This cutaway diagram of the Space Shuttle orbiter reveals a huge payload bay which can carry a variety of payloads. The crew spend most of their time in the pressurized nose of the craft.

States

PAYLOAD BAY 60 x 15 FT, 60 FT BAY DOORS, IT CAN CARRY 29 T INTO ORBIT

DANGERS OF SPACE

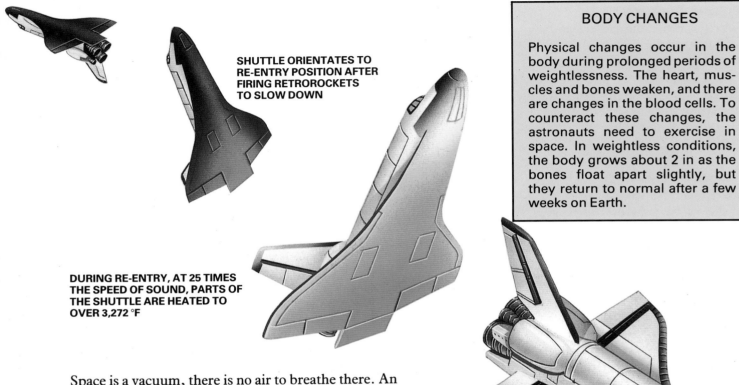

SHUTTLE ORIENTATES TO RE-ENTRY POSITION AFTER FIRING RETROROCKETS TO SLOW DOWN

DURING RE-ENTRY, AT 25 TIMES THE SPEED OF SOUND, PARTS OF THE SHUTTLE ARE HEATED TO OVER 3,272 °F

BODY CHANGES

Physical changes occur in the body during prolonged periods of weightlessness. The heart, muscles and bones weaken, and there are changes in the blood cells. To counteract these changes, the astronauts need to exercise in space. In weightless conditions, the body grows about 2 in as the bones float apart slightly, but they return to normal after a few weeks on Earth.

THE SHUTTLE IS A GLIDER DESCENDING SEVEN TIMES STEEPER AND 20 TIMES FASTER THAN AN AIRPLANE

Space is a vacuum, there is no air to breathe there. An astronaut needs a pressurized cockpit and, to work outside the spacecraft, a spacesuit. Astronauts also need to be protected from the intense heat of the sunlight and the intense cold of the darkness in space.

Van Allen Belts

The Earth is surrounded by two radiation belts. These are made up of atomic particles from the Sun, trapped by the Earth's magnetic field. Called Van Allen belts after the scientist who discovered them, they lie about 2,000 and 13,000 mi above the Earth. It is dangerous to venture for long at altitudes in which these radiation belts exist.

Space Pollution

Man has started to pollute space, creating a new kind of hazard as millions of man-made objects orbit the Earth in every direction. This debris ranges from flakes of paint and insulation shed by spacecraft to discarded satellites which no longer work and the rocket stages used to launch them. A flake of paint traveling in the opposite direction to a Space Shuttle, with a closing speed of 35,000 mi/h, could smash the window, depressurizing the craft and killing the crew.

Above: The Shuttle lands like an airplane after entry into the Earth's atmosphere. Friction heats the Shuttle to over 3,200 °F so the aluminum structure is protected by special heatshield tiles, the nose cap, and leading edges of the wings by an extremely strong carbon material.

PART OF SKYLAB 6.5 X 3.2 FT FELL TO AUSTRALIA • 22 LB PIECE OF SPUTNIK FELL TO USA

Space Travel

Getting into and out of space is dangerous. The rocket can blow up during launch when the propellents are ignited. As the launch rocket accelerates, the astronauts suffer a tremendous gravitational pull on their bodies. These g-forces make them feel very heavy. Once in space, the astronauts become weightless and float inside the spacecraft. Without strenuous exercise and careful health checks, long journeys in weightless conditions can be dangerous because the body has to withstand the strain of returning to gravity. On re-entering the Earth's atmosphere, g-forces again pull on the body, then, after the spacecraft lands, Earth's continuous one-g gravity can have detrimental effects on a body that has been used to weightlessness for very long periods.

The re-entry itself is very hazardous. Traveling at 18,000 mi/h, the spacecraft is heated up as it plunges into the outer layers of the Earth's atmosphere. Friction with the air causes the spacecraft to glow and, without a heatshield, it would burn up. Fortunately, there has never been a heatshield failure on a manned spaceflight – yet.

Right: A US Shuttle astronaut wears a spacesuit with a portable life support system backpack to work outside the craft.

LANDING AT 250 MI/H

SPACE SHUTTLE RE-ENTERS ATMOSPHERE AT 18,000 MI/H, HEATED TO OVER 3,200 °F

SPACE STATIONS

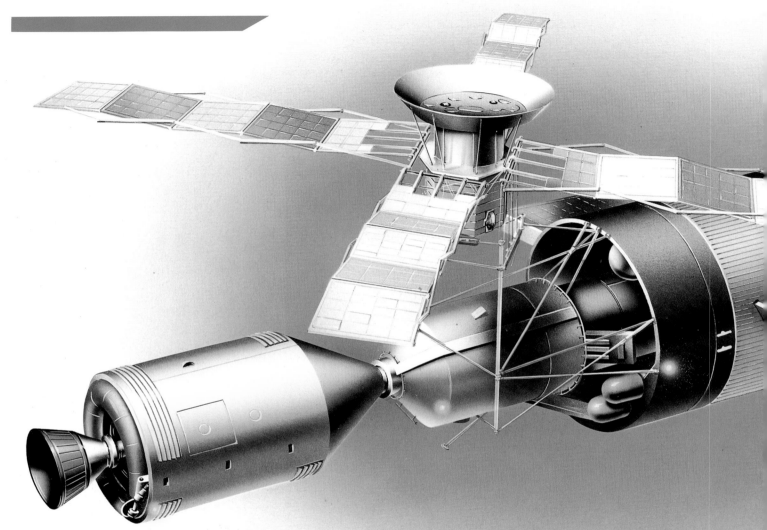

In science fiction books and films, space stations are usually crewed by hundreds of people living in a self-supporting environment. The reality is very different. Space stations are orbiting spacecraft in which a few astronauts stay for weeks or months at a time to carry out scientific experiments.

US Space Stations

In 1973, the Americans launched a space station called Skylab. This huge craft consisted of the top stage of a Saturn V rocket which was converted to carry American crews of three at a time. Skylab contained equipment for carrying out scientific experiments, such as processing materials, and observing the Earth and Sun. The astronauts also carried out medical experiments on themselves to study the effects of prolonged spaceflight on the human body.

The Skylab project finished in 1974 but, since then, America has not launched another station. There are plans for a new space station called Freedom, to be built in the 1990s, but these are threatened with budget cuts. If Freedom is built, it will consist of several pressurized, cylindrical modules providing living quarters and laboratories, joined together and mounted on a giant girder. The space station will be equipped with large solar panels to provide electricity.

Soviet Space Stations

The USSR has taken the lead in space station flights. Soviet space stations are basically cylinders as big as a single railway carriage with wings. The first, called Salyut 1, was launched in 1971 and was 46 ft long. A new space station called Mir was launched in 1986. It is an improved Salyut cylinder, to which the Soviets

MIR (MEANING "PEACE") LAUNCHED BY USSR IN 1986, 44 FT LONG, WEIGHT 20 T

Left: The US Skylab space station was launched in 1973 to house three astronauts. It re-entered the Earth's atmosphere in 1979.

are adding new cylindrical modules so that eventually the Mir complex will consist of five modules joined together; each module will be used for different purposes. Plans have been thwarted by technical problems and budget cuts. By 1990, when the complex was supposed to have been completed, only three new scientific modules, called Kvant, had been added.

The Soviets have been able to fly crews for missions lasting one year and may fly longer missions; fresh supplies and equipment are ferried up by an unmanned tanker called Progress, which is controlled from the ground. Seven cosmonauts have amassed over 300 days' spaceflight experience, and one, Yuri Romanenko, has clocked up 430 days on three flights. The longest spaceflight is 365 days by Vladimir Titov and Musa Manarov in 1987–88.

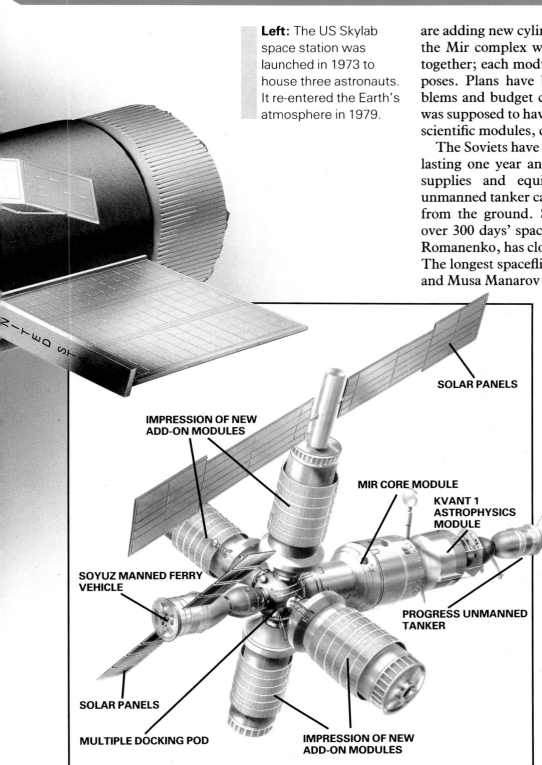

SOLAR PANELS

IMPRESSION OF NEW ADD-ON MODULES

MIR CORE MODULE

KVANT 1 ASTROPHYSICS MODULE

SOYUZ MANNED FERRY VEHICLE

PROGRESS UNMANNED TANKER

SOLAR PANELS

MULTIPLE DOCKING POD

IMPRESSION OF NEW ADD-ON MODULES

SPACE STATIONS

Salyut 1 (1971, USSR): three cosmonauts stay for record 23 days (but are killed in a re-entry accident).
Skylab (1973, USA): three crews of three astronauts each stay for record 28-, 59- and 84-day missions.
Salyut 3 (1974, USSR): first military space base.
Salyut 4 (1974, USSR): Soviets stay for 62 days.
Salyut 5 (1976, USSR): military space base
Salyut 6 (1977, USSR): crews stay for 96, 139, 175 and 184 days.
Salyut 7 (1982, USSR): flights extended to 211 and 236 days.
Mir (1986, USSR): first flight to exceed 300 days, then record 365 days.

Left: An artist's impression of the completed Soviet Mir space station complex.

LIVING IN SPACE

The first problem astronauts encounter when they enter space is weightlessness because of the lack of gravity. Floating around a spacecraft makes it difficult for them to work and carry out experiments, and it can make some space travelers feel sick. About half of those who travel into space feel ill for the first few days before they acclimatize to the weightlessness.

On very long journeys, it is absolutely essential that two hours each day are spent rigorously exercising. Otherwise the muscles and heart would soon weaken as they have little work to do in zero-gravity. On the Space Shuttle, astronauts jog on a treadmill wearing an elasticated harness to hold them down.

Eating, sleeping and going to the bathroom are much the same as on Earth except that weightlessness has to be taken into account. Food is in small containers, so that it can be scooped out with a spoon. Most of the food is moist so it does not break up into crumbs that float off in the weightlessness. To sleep, the astronauts zip themselves inside sleeping bags which are attached to the bunks. The toilet has a special vacuum flush to suck the waste into a tank.

Suiting Up

While the astronauts are inside the pressurized spacecraft, they can wear casual clothes for comfort, but to work outside the spacecraft, they need space-suits. These not only provide them with oxygen under pressure, but also give protection from the dangers outside, such as the extreme heat or cold, the vacuum (lack of air), the radiation, and the possible impact of micrometeorites or small pieces of space debris. Space suits are made of several layers of insulating material, with a white outer covering to reflect the Sun's rays. Under this is underwear which has tubes of water running through it to cool the astronaut. A backpack, called the portable life support system, carries oxygen and cooling water for the underwear, as well as a radio unit.

One of the newest space inventions is the manned maneuvering unit, which enables an astronaut to fly independently using gas thrusters on the unit to move about in space. Otherwise, astronauts have to wear a safety harness when working outside the spacecraft to prevent them floating away into space.

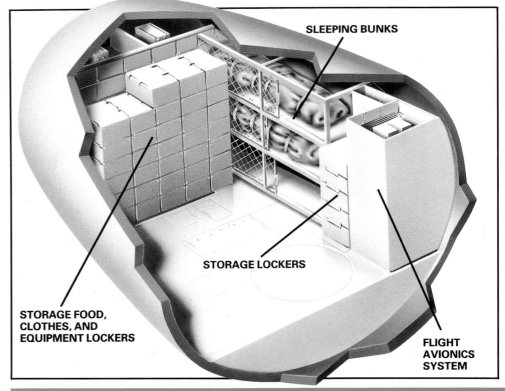

SLEEPING BUNKS

STORAGE LOCKERS

STORAGE FOOD, CLOTHES, AND EQUIPMENT LOCKERS

FLIGHT AVIONICS SYSTEM

Left: Even on the Space Shuttle, living space is cramped. The lower mid-deck has to serve as a locker room, bedroom, and bathroom.

Right: The first person to become an independent satellite, Bruce McCandless. In 1984, using a hand-controlled device called a manned maneuvering unit, he flew 300 ft from the Shuttle.

LONGEST MANNED SPACEFLIGHT 365 DAYS BY VLADIMIR TITOV AND MUSA MANAROV

LARGEST CREW IN SPACE IS EIGHT, ON *CHALLENGER 9* SPACE SHUTTLE, IN 1985

LIFE AMONG THE STARS

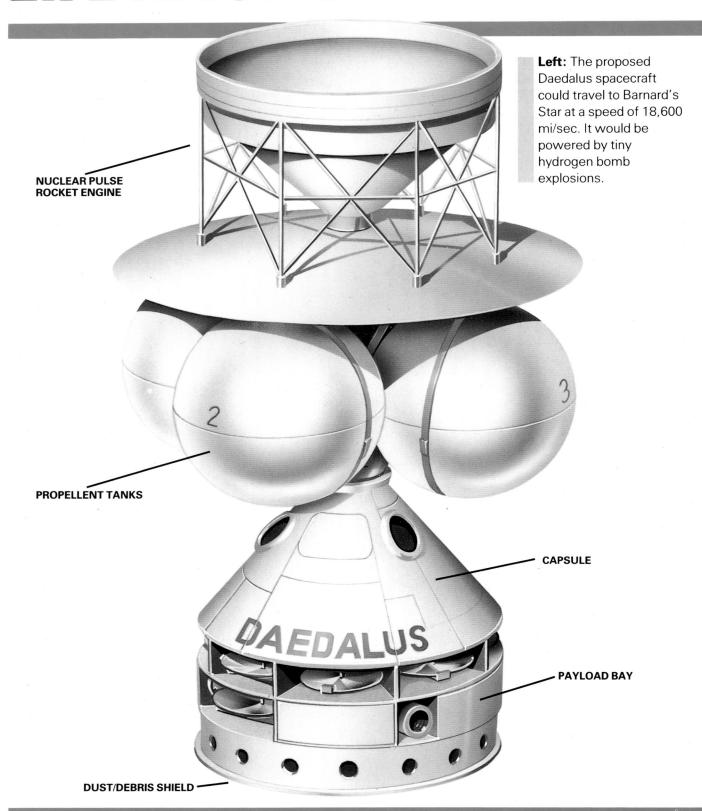

**NUCLEAR PULSE
ROCKET ENGINE**

Left: The proposed Daedalus spacecraft could travel to Barnard's Star at a speed of 18,600 mi/sec. It would be powered by tiny hydrogen bomb explosions.

PROPELLENT TANKS

CAPSULE

PAYLOAD BAY

DUST/DEBRIS SHIELD

DAEDALUS, THE FIRST STARSHIP, WILL TRAVEL AT 18,600 MI/SEC TO BARNARD'S STAR

In science fiction, spacecraft travel to planets inhabited by strange beings and establish space stations deep in space. But to do this, our spacecraft need to travel faster. The fastest spacecraft we have sent into space is Voyager 2; in 12 years it has traveled 3,000 million mi. It is now on course for Sirius, which is the brightest star in our sky, and also one of the closest stars to Earth, but even so is still 8.64 light years away. Voyager will pass it at a distance of 0.8 light years in just 368,000 years time!

Daedalus – The First Starship

The British Interplanetary Society have designed a starship called Daedalus. This unmanned probe will be sent on a 50-year trip to Barnard's Star. Only 5.9 light years away, it is one of the closest stars and seems to have a planetary system around it. Daedalus is 20 times the size of the Apollo Saturn V rocket and will be powered by a nuclear fission engine which will enable it to travel at 0.1 times the speed of light – 18,600 mi/sec.

After reaching Barnard's Star, Daedalus will transmit data from the 450-t monitoring equipment and sensors. The spacecraft will have an enormous nuclear generator to create the power needed to send strong radio signals. These may be picked up on Earth six years later.

Searching for Other Life

Mankind has two methods of searching for possible other life in space: sending spaceprobes to the planets of our Solar System and listening with radio telescopes for possible messages from distant civilizations.

The Voyager 2 probe is carrying messages to any civilization that may one day interrupt it. In the grooves of a record are encoded written messages, pictures of our planet, spoken greetings, various sounds of life on our planet, plus some of the Earth's greatest musical hits.

Messages have also been sent into space by radio telescope. In 1974, the Arecibo radio telescope in Puerto Rico sent a message in code to a distant star cluster called M13 in the constellation of Hercules. This message from the world's largest radio telescope told aliens about our body chemistry, appearance and size, and our Solar System. The star cluster is so far away that even if any aliens there replied immediately,

we would not hear from them until AD 50000.

Some people say they have seen UFOs or unidentified flying objects, but these turn out to be bright planets and stars, aircraft, meteors and satellites.

Some people believe that God created "the Heaven and the Earth" and made life on Earth and man "in his own image". The Bible doesn't mention life on any other world in space. How do we know that there are other worlds in space capable of sustaining life? There are nine planets orbiting our Sun and none of the others can support life as we know it. Finding other worlds that may be capable of supporting lifeforms are so far away that it would take thousands of years to make contact. Today, when so many man-made satellites are constantly circling the earth, we can hardly believe that any creatures from outer space could land on Earth and leave without detection. For the time being, we must be content to look at the beautiful photographs of Earth taken by the Apollo astronauts. We need to take care of our Earth far more than we need to search for life elsewhere.

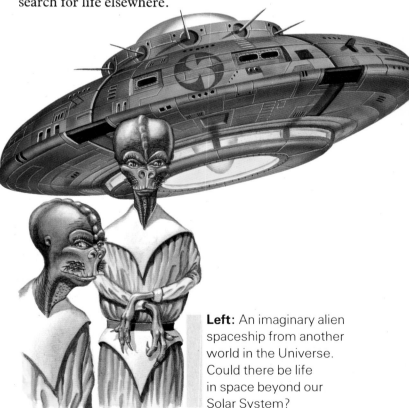

Left: An imaginary alien spaceship from another world in the Universe. Could there be life in space beyond our Solar System?

WE KNOW THERE IS NO OTHER LIFE IN OUR SOLAR SYSTEM, BUT BEYOND THIS?

INDEX

ACKNOWLEDGMENTS

The Publishers wish to thank the following photographers and organizations for their kind permission to reproduce their photographs in this publication:

Amateur Astronomy 48; Associated Press/Topham 30, 44 top, 63 top; Daily Telegraph Colour Library 6, 7, 8–9, 20, 21, 28 below, 51, 54, 63 below, 67; Octopus Group Picture Library/NASA 64, 65 /Royal Astronomical Society 14 below; Science Picture Library 15, 29 /Julian Baum 45 /European Space Agency 61 below /NASA 11, 14 top, 32, 42, 43, 44 below, 53, 56–57 /Novosti 31 /David Parker 15 /US Department of Defense 62; NASA 28 top, 34, 35, 39, 61 top, 71, 75; Topham 33, 41

Illustrations by:

David Bergen – pages 12–13, 21 top, 33, 36–37, 39, 46–47, 53 top, 64, 66, 77

The Maltings Partnership – pages 6–7, 8, 11, 17 top, 18–19, 21, 22, 23, 25 top, 26–27, 31, 35 top, 37 top, 39 top, 40–41, 43 top, 45 top, 46, 50–51, 58–59, 63 top, 67 top, 70–71, 75 top, 76

Oxford Illustrators – 16–17, 24–25

Illustrations in the pull-out time chart show, from left to right:

Isaac Newton and his Reflecting Telescope; The Moon; Jodrell Bank Radio Telescope, England; Venera 4; Jupiter; Halley's Comet; Mariner 10; Saturn; Robert Goddard and his rocket; Lunar landing; The Space Shuttle Columbia; Yuri Gagarin; Skylab; Bruce McCandless.

1959 Soviet Luna 2 hits Moon

1959 Luna 3 photographs far side of Moon

1962 First successful spaceprobe, Mariner 2, flies past Venus

1964 First close-up photographs of Moon

1965 Mariner 4 sends back close-up photographs of Mars

1966 Luna 9 soft lands on Moon

1966 Luna 10 orbits Moon

1966 Lunar orbiter begins Moon mapping program

1967 Venera 4 explores atmosphere of Venus

1968 Apollo 8 orbits the Moon with three astronauts

1969 First men land on Moon in Apollo 11

1970 Venera 7 is first spacecraft to land on Venus and return data

1970 Luna 16 returns samples of Moon to Earth

1970 Luna 17 deploys unmanned lunar rover

1971 Apollo 15 astronauts drive lunar rover

1971 Mariner 9 is first spaceship to enter Mars' orbit

1972 Apollo 17 makes the final manned flight to the Moon

1973 Pioneer 10 explores Jupiter

1974 Pioneer 11 explores Saturn

1974 Mars 5 becomes only successful Soviet Mars probe

1974 Mariner 10 is first spaceprobe to explore Mercury

1975 Russian Venera 9 and Venera 10 soft land on Venus and send back first pictures from surface

1976 Vikings 1 and 2 make the first soft landings on Mars

1978 Pioneer 12 orbits Venus and deposits landers

1979 Voyager 2 flies past Jupiter

1980 Voyager 1 flies past Saturn, instruments show complex ring system

1981 Voyager 2 flies past Saturn

1982 Venera 13 returns first color pictures from surface of Venus

1985 Vega 1 and 2 deploy balloons into atmosphere of Venus

1986 Armada of spacecraft, led by Europe's Giotto, explores Halley's Comet

1986 Voyager 2 flies by Uranus

1989 Voyager 2 flies by Neptune

1990 Japanese place lunar probe in orbit.

1990 Galileo en route to enter orbit around Jupiter and deploy atmosphere probe

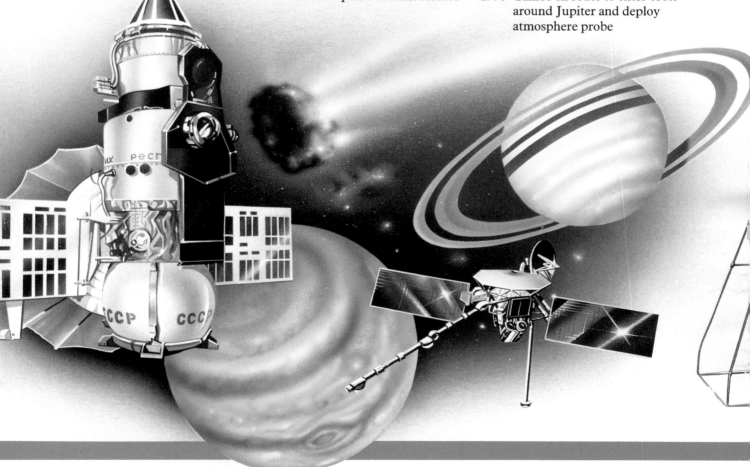

1232 China launches first rockets

1806 British soldier William Congreve launches large solid propellent rocket in battle against Boulogne

1903 Russian schoolteacher Konstantin Tsiolkovsy foresees manned space travel

1926 American physicist Robert Goddard launches first liquid propellent rocket

1943 Germany launches V2 rocket

1951 Viking 7 rocket reaches height of 124 mi

1955 USA and USSR propose launches of satellites

1957 USSR launches first intercontinental ballistic missile

1957 USSR launches first satellite, Sputnik 1

1957 USSR launches dog, Laika, into orbit

1958 America launches satellite

1959 Luna 2 hits Moon

1959 First spy satellites launched

1959 Luna 3 photographs far side of Moon

1960 USA launches first weather and navigation satellites

1960 Capsule recovered from orbit by USA

1960 USSR recovers two dogs, Belka and Strelka, from orbit

1961 Yuri Gagarin in Vostok 1 is the first man to be launched into space

1961 President John F Kennedy begins Project Apollo to put men on Moon

1962 First communications satellite, Telstar, launched

1964 Syncom 3 is the first communications satellite to be launched into geostationary orbit over the Pacific Ocean

1965 First rendezvous in space between Gemini 6, and Gemini 7, already in orbit

1966 Gemini 8 made first docking in space

1967 First flight of Saturn V booster

1968 First men orbit Moon in Apollo 8

1969 Manned lunar landing by Apollo 11

1971 USSR launches first space station, Salyut 1

1972 Pioneer 10 launched on mission to leave Solar System

1972 First Earth monitoring satellite launched

1977 Space Shuttle glide tests in atmosphere

1979 Europe launches first Ariane satellite booster

1981 First flight by Space Shuttle with Columbia orbiter

1984 First retrieval of satellites in space by Shuttle

1986 Space Shuttle Challenger explodes shortly after launch

1987 USSR launches massive Energia booster

1988 First manned spaceflight to last one year aboard Mir space station

1988 USSR launches unmanned Space Shuttle, Buran

1989 First direct broadcast TV satellites launched

TIME CHART ASTRONOMY

BC

4–3000 Chinese observe eclipses of Moon and Sun

3000 Egyptians develop astronomy as a science

600 Greek philosopher, Thales of Miletus, considers Earth may be round

500 Greeks understand phases of the Moon

400 Greek astronomer Eudoxus says Earth rotates

300 Aristarchus says Earth orbits Sun

200 Hipparchus compiles star catalog of over 1,000 stars

AD

200 Ptolemy produces a revolutionary astronomy encyclopedia

827 Ptolemy's encyclopedia is translated into Arabic as the *Almagest*

1054 Chinese observe a supernova where the Crab nebula is today

1543 Nicolaus Copernicus says Sun is the center of the Universe

1608 First telescope designed by Hans Lippershey

1609 First known observations using a telescope made by Galileo

1609 Johannes Kepler says that planets orbit the Sun

1666 Isaac Newton formulates laws of gravity

1668 Newton builds first reflecting telescope

1675 Speed of light measured as 186,282 mi/sec

1705 Edmond Halley predicts a comet will appear in 1758. It does, and is named after him

1767 *Nautical Almanac* of heavenly bodies published

1781 Charles Messier's catalog of nebulas and star clusters published

1781 William Herschel discovers the planet Uranus

1801 Giuseppe Piazzi discovers first asteroid, Ceres

1802 William Herschel discovers binary stars

1838 Friedrich Wilhelm Bessel measures the distance to the first star: 61 Cygni

1846 Eighth planet, Neptune, first predicted independently by John Couch Adams and Urbain Leverrier, subsequently discovered by Johann Gottfried Galle

1863 Pietro Angelo Secchi classifies stars by spectral type

1877 Giovanni Schiaparelli observes "canals" on Mars

1905 Albert Einstein announces the Theory of Relativity

1918 100-in reflecting telescope built on Mt Wilson, USA

1925 Edwin Powell Hubble uses 100-in telescope to show other galaxies are receding from us, confirming idea of expanding Universe

1930 Pluto discovered by Tombaugh

1931 Karl Jansky discovers radio waves from outer space

1937 First radio telescope built by Grote Reber

1948 200-in Palomar reflecting telescope built

1955 Jodrell Bank radio telescope built

1963 Quasars first detected

1967 Pulsars discovered